Digital Sensations

Electronic Mediations

Katherine Hayles, Mark Poster, and Samuel Weber, series editors

Electronic Mediations, volume 1

Digital Sensations

Space, Identity, and Embodiment in Virtual Reality

Ken Hillis

 University of Minnesota Press
Minneapolis • London

The University of Minnesota Press gratefully acknowledges permission to reprint the following. An earlier version of chapter 1 appeared as "A Geography of the Eye," in *Cultures of Internet*, edited by Rob Shields, copyright 1996, Sage Publications, Ltd., and the author. Portions of chapter 2 appeared as "Toward the Light 'Within': Optical Technologies, Spatial Metaphors, and Changing Subjectivities," in *Virtual Geographies*, edited by Mike Crang, Phil Crang, and Jon May, copyright 1999, Routledge. Earlier versions of portions of chapters 5 and 6 appeared as "Human.language.machine," in *Places through the Body*, edited by Heidi Nast and Steve Pile, copyright 1998, Routledge.

Published by the University of Minnesota Press
111 Third Avenue South, Suite 290
Minneapolis, MN 55401-2520
http://www.upress.umn.edu

Library of Congress Cataloging-in-Publication Data

Hillis, Ken.
    Digital sensations : space, identity, and embodiment in virtual reality / Ken Hillis.
        p.   cm. — (Electronic mediations ; v. 1)
    Includes bibliographical references and index.
    ISBN 0-8166-3250-2 (hc). — ISBN 0-8166-3251-0 (pb)
    1. Human-computer interaction.   2. Virtual reality.   I. Title.
    II. Series.
    QA76.9.I58H53   1999
    006—dc21   99-32322

Printed in the United States of America on acid-free paper

The University of Minnesota Press is an equal-opportunity educator and employer.

11 10 09 08 07 06 05 04 03 02 01 00 99   10 9 8 7 6 5 4 3 2 1

Contents

Acknowledgments

This book is organized around the *ideas*—the philosophies, discourses, political assumptions, and, yes, even the magical thinking—that inform Virtual Reality. I think of this technology as resulting from, at least in part, the age-old desire to control the natural world. And if the virtual environments that Virtual Reality permits are not exactly, to borrow a line from Tennyson, "red in tooth and claw," as a geographer of communication, I am fascinated by how they appropriate from, and comment on, some of the ways that we think about nature, space, and the places in the world that we make. Virtual Reality is a communication technology that relies on images of space and place within which and with which its users interact. It supports the fantasy that communication of messages or information, and the conduits through which they are transmitted, together might offer an adequate imaginary space that would substitute for aspects of the material world considered by many to be exhausted, in retreat, inaccessible, or too limited and constraining for an imagination that yearns to be free of its human body or "home base."

How do we know what we know? How do we organize and communicate this knowledge? How might a form of knowing, for better or worse or both, intersect with the places we make? How does communication technology influence the ways we transform and use the nonhuman parts of the natural world? These are ancient yet still vital questions, and I hope this work makes a modest contribution toward furthering their discussion. Today such discussion also demands bringing to the table issues of visual culture, and visual communication, and a recognition of the need for new ways to think about human relationships to the natural world in face of the power of optical technologies of representation. I also intend the book as a political and ethical counterweight to the promotional hype that subtends research and development of Virtual Reality yet also occludes recognition of the wider history and geography of vision within which the technology is situated. As an idea and a material practice, Virtual Reality is produced within a culture of increasing surveillance and voyeurism, whose members often are camera ready and frequently "like to watch."

Digital Sensations is the outcome of my Ph.D. dissertation in human geography, written at the University of Wisconsin–Madison. Thanks are due to the members of my committee — Bob Sack, Yi-Fu Tuan, David Woodward, John Fiske, and Lew Friedland — all of whom offered valuable criticism and direction to this seeker of truth in the information age. Gerry Kearns's scholarship and wit were sustaining at difficult moments. The thoughtful, detailed, and caring reviews offered me by readers of the manuscript allowed me to sharpen and advance the arguments herein enormously. Thank you Helen Couclelis, Martin Jay, Steven Shaviro, and Rob Shields. During my hiatus at the University of Colorado–Boulder, Don Mitchell offered a much-needed ear. His invitation to present a synopsis of this research as part of the Department of Geography's Colloquium Series afforded me a most useful discussion with audience members. Jon May's incisive and supportive critique of portions of two chapters allowed me to strengthen the arguments therein.

I owe particular gratitude to Michael Petit, whose critique, editorial commentary, and partnership have sustained me throughout the book's production and proved pivotal to its completion. Paul Couillard read the entire manuscript at the dissertation stage, and no doubt he will recognize how his insights have contributed to these pages. Liz McKenzie is a friend and artist. She and I have spent many productive hours discussing the relationships between the images in figure 6, for which she provided the artwork, and the written descriptions of the ideas they model. Carrie Mullen's guidance and editorial support throughout the process have been invaluable. So has that of Robin A. Moir, Laura Westlund, and Bill Henry. My research assistant at the University of North Carolina, Renee Lertzman, lightened the load of final research and production details. Thanks are also due in particular to Scott Luscombe, and to Apurva Uniyal, Jeff Zimmerman, Karen Till, QingLing Wang, Jonathan Perry, Drew Ross, Paul Adams, Altha Cravey, Leo Zonn, and Bart Nagel.

I must also thank my colleagues at the Department of Communication Studies, the University of North Carolina at Chapel Hill, for their encouragement and support. Bill Balthrop, chair of the department, has been exceptionally supportive of my research interests. The final stages of research and writing were supported by funding from the university's Institute for Research in the Social Sciences, as well as a UNC Junior Faculty Development Award. The Social Sciences and Humanities Research Council of Canada provided financial support for research and writing of the dissertation.

List of Abbreviations

AI	artificial intelligence
ARL	U.S. Army Research Laboratory
ARPA	Advanced Research Projects Agency
CAD	computer-aided design
CBD	central business district
CRT	cathode-ray tube
ENIAC	Electronic Numerical Integrator and Computer
GIS	Geographic Information Systems
HITL	Human Interface Technology Lab
HMD	head-mounted display
IRC	Internet Relay Chat
IT	information technology
MUDs	multi-user dungeons (or domains)
NASA	National Aeronautics and Space Administration
PC	personal computer
SF	science fiction
VE	virtual environment
VR	Virtual Reality

Introduction: Digital Relations

But of all the sciences Optics is the most fertile in marvelous expedients.
Sir Daniel Brewster, *Letters on Natural Magic*

The world conveyed by the interactive computer has been dubbed "virtual" because its location or features cannot be pinpointed in the tangible world. It exists within the relation between the machine and the user. We cannot place it inside the machine, because it is not there unless we invoke it, and it is not wholly within our minds because we do not possess the hardware necessary to conjure it up.... In the computer... we can move throughout a constructed universe of our own making, on virtual paths invisible even as we tread upon them.
David Rothenberg, *Hand's End: Technology and the Limits of Nature*

Over the last few years, Virtual Reality, or "VR," has become something of a household term. Discussion in the popular media abounds, and a number of speculative, promotional books on the subject have achieved mass-market success. Promotional writing is a part of the hype surrounding VR. Barrie Sherman and Philip Judkins, for example, find that the technology and the experiences VR affords are "a proxy for the American dream — to be at the centre, the President, a star in your own Hollywood movie" (1993, 29). Despite VR being accorded the hype of celebrity status — facilitated in part by cultural fictions such as *Max Headroom, VR5, Johnny Mnemonic,* and perhaps most importantly the holodeck on *Star Trek: The Next Generation* — a lack of theorization exists that would provide greater understanding about why VR, both as a technology and as an idea, has emerged at this cultural moment. This book is an effort to help close that gap. A critical approach to VR is crucial at a time of a widespread rush to laud its liberatory possibilities and thereby accord VR great "cultural capital" on the part of media and publics, many academics included.

VR is a technological reproduction of the process of perceiving the real, and I am interested in examining how representational forms generated within communications technologies, especially digital virtual technologies, affect "the lived world."[1] Today, our lived worlds are plural, inflected by conceptions of space and time specifically segregated from one another. Segregated spaces and times require means to communicate among them; their mutual compartmentalization enhances and extends a sense of distance among our various lived worlds, thereby abetting a wide cultural acceptance of communications and information technologies, or IT, as necessary and natural. One focus of my project examines how theories of absolute, relative, and relational space are incorporated into virtual technologies, which themselves are the material basis permitting the imaginary, digital dataspaces popularly called Virtual Reality to take form. These metaphoric spaces challenge distinctions among different concepts of space and weaken distinctions between geographic conceptions of place and of landscape.

As a term, "VR" is denigrated by many industry players yet is used with relish by other industry and academic spokespeople. The popular media consider the term adequate — referring as it does to both immersive and nonimmersive forms of interactive communication technologies and experiences. "VR" is often used interchangeably with the catchall term "cyberspace" to encompass the VR experience. Marcos Novak idealizes cyberspace as a

> spatialized visualization of all information in global information processing systems, along pathways provided by present and future communications networks, enabling full copresence and interaction of multiple users, allowing input and output from and to the human sensorium, permitting simulations of real and virtual realities, remote data collection and control through telepresence, and total integration and intercommunication with a full range of intelligent products and environments in real space. (1992, 225)

Because of its cultural popularity, the term "VR" is used throughout this book. Wherever possible, however, I distinguish between virtual technologies and the 3-D computer-generated virtual environments (VEs) they permit. As spatialized realms of digitally coded information, VEs are made possible by virtual technologies. They use iconographics, which are more conducive to collapsing experiential differences and distances between symbols and referents, or the virtual and the real, than text-based applications such as E-mail, Internet listservs, bulletin boards (BBSs), newsgroups, or Internet Relay Chat (IRC).[2]

I distinguish between virtual technologies and VEs not because the term "VR" is "incorrect" or imprecise per se. "VR" is a hybrid term. It refers to an individual experience constituted within technology, and it draws together the world of technology and its ability to represent nature, with the broad and overlapping spheres of social relations and meaning. Modern epistemologies and academic disciplines have worked very hard, sometimes unwittingly, to hold these spheres apart. Because this gap or distinction between technology and social relations and meaning is made to seem natural, it is productive to distinguish between technical and social components in theorizing VR as a postmodern technology, practice, and idea.

I consider VEs to be representational spaces that propose particular spatial illusions or fantasies, and I understand "Virtual Reality" to constitute what might more profitably be termed a virtual geography. In addition to describing these technologies and environments in a way that historicizes and theorizes them, I develop tools to analyze these geographic fantasies and the substantive experiences they allow. An immersive technology such as VR is a product at once both sedative and stimulating. It breaks with yet extends modernist techniques and conventions. Therefore, I examine how VR the technology might influence contemporary self-perception of our identities — both imaginative and embodied — and the places we inhabit. Hubert Dreyfus (1992) argues that the West has a penchant for turning its philosophical assumptions into technologies. With VR, one assumption being transformed from idea to action is that a series of extant social relations based on an individualistic understanding and practice of pluralism might be relocated to a disembodied datascape — an immaterial landscape "wherein" military exercises, commercial transactions, virtual "on-the-job" training, and so on increasingly "take place." This relocation of what is concrete to an imaginary or metaphoric space assumes that the act of communication is a wholly adequate substitute for embodied experiential reality; it exchanges communications technologies for the reality of places and dispenses with, for example, empiricism's concerns about sense data and how things are understood as true and/or real. This relocation to virtual spaces also relies on a second assumption whose history at the very least parallels René Descartes's dictum "Cogito ergo sum." As mere automata, our animalistic and all too finite physical bodies are thought secondary to our minds and representational forms—a dynamic that is built in to virtual technologies. As such, VR exemplifies a specific postmodern irony—or, at least, a distinctly modern paradox: as a practice and spatial representation, it reveals an unacknowledged belief in a hybrid that I term "magical empiricism."

Theorizing human bodies as secondary to the ways they are represented has a cultural history, and VR has developed within specific cultural contexts. Generally speaking, culture—as a range of material practices and technical and intellectual works, also reflected in individual and collective ideas, desires, and aspirations—can operate to shield us from the brute reality of certain aspects of our embodiment. Culture offers an ironic form of security that denies the real. To varying degrees, and partly depending on how the powerful operationalize the *notion* of culture, all cultures facilitate this "escape" from the body and its needs and actions involving food, sex, and death. Escape can involve a denial of the constraints of embodied reality. It can also take the form of a flight from the reality of "the other," from an oppressive set of dominant social mores, political expectations, and restrictions. At an earlier time in the United States, for example, individuals and entire communities—easterners operating within an experiential mode of cultural disaffection with, and/or a political and religious alienation from, the "here and now"; Midwesterners finding that Thomas Jefferson's enlightened merger of agrarian life and individual happiness was incoherent and incompatible with their goals for a comfortable, even paradisiacal existence; all seekers in hopeful anticipation of the future "elsewhere"—uprooted themselves and pushed ever westward toward the geographic frontier. If this frontier has been "closed" for a century, the desire to escape from contingency has not diminished. It is more than interesting that the impetus toward relocation into virtual living is most pronounced in California—that earlier real frontier filled with emigrants who believed they had arrived in Utopia and left history behind. However, I would also note that physical movement across the space between "here" and "there" is intimately associated with American utopian connections among individualism, freedom, and space conceived more in terms of extension than in terms of engagement. Route 66 is now a memory—a set of historical markers and asphalt traces, a 1960s TV program, and mugs and sweatshirts available from souvenir outlets along U.S. Interstate 40 in New Mexico. If an escapist movement in real space toward an unpopulated and virgin promised land is now problematic or unavailable, for many, seeking out and creating "information superhighways" that permit "migration" to new "electronic frontiers" offers an imaginative and apparently compelling utopian alternative to physically going "on the road."

In the earlier American trek westward, an environmental determinism was at work. A utopian ideal of purity was projected onto a frontier nature uncontaminated by European cultural influences, yet awaiting cultural conquest by a purportedly more noble, rugged, and enlightened "new world spirit." This utopian ideal offered moral uplift and formed part of a synergy by reflecting in equal measure the self-presumed morally elevated spirit of those choosing to make the trek. Reflecting an updated belief that escape from aspects of the real is possible through spatial relocation, cyberspace and VR are today's utopian "feedback controlling machines"[3] — updated models for imagining utopias that promise not only renewal and deliverance but also rational social control over space, information, and identity. Immersive virtual environments can be thought of as a form of cosmographic mapping, and within this understanding, cyberspace and VR are, respectively, a frontier metaphor and a technology offering both the promise of an escape from history with a capital *H*, and the encrusted meanings it contains, and an imaginary space whereby to perform, and thereby possibly exorcise or master, difficult — even contradictory — real-world historical and material situations. Cyberspace and VR promoters and enthusiasts, however, tend to deny the meanings, contexts, social relations, and political implications that inform and attend the move to virtual living — a move of which I remain skeptical. Cyberspace not only suggests that an ideal existence is one that is technologically mediated; it also continues and intensifies a long-standing project to alter, via the use of technology, subjectivity and the meaning of what it is to be human (see Kendrick 1996).

A further irony is worth noting. Westward migration involved a conquering, taming, or pushing back of nature, but it also often meant a hands-on engagement with the natural world by those who had left a relatively more urbanized environment. If culture offers a way to deny the contingency of the real, as a part of culture, technology is now being positioned to suggest an alternative to the natural world. Increasingly, however, technology is a constant mediating force in everyday life. Yet it is suggested that VR will free our imaginations from mundane realities themselves inflected by technology. Further, as a technology, VR becomes a narrative of the future that abets forgetting the failed promises of earlier technologies.

Consider for a moment that VR is being positioned as a place where many may go to experience a natural world repositioned as a luxury commodity unavailable to all but the most well-heeled. Stuart Aitken and Andrea Westersund (1996) at San Diego State University are in the process of documenting how the Catalina Conservancy, which manages the wild portions of Santa Catalina Island off the coast of southern California, is constructing a virtual environment that would allow tourists to visit threatened natural ecologies on the island without actually going there. This would allow threatened areas time to heal from the wounds of overvisitation, yet as part of the same proposal, the Conservancy is also promoting jeep eco-tours of the island. You and I will tour Catalina courtesy of the head-mounted display (HMD). However, for $795.00 per day, a "naturalist-trained" driver will safely transport well-heeled eco-tourists through the real thing now off-limits to the hoi polloi. Virtual tourists will have to deal with the implicit suggestion that virtual space may be better than the "real thing"; however, for the wealthier, VR also confirms that "real" remains best and merits the higher price of fully embodied admission.

In other educational settings, students are told, for example, that they will not need to visit places such as the Peruvian Andes. Instead, they will enter a VE simulation of this far-off reality and, by engaging with a series of interactive images, obtain an experience as good as being there. As figure 1 depicts, in this future fantasy, those who seek nature will now join with the technology in a kind of ecstatic, out-of-body, New Age reunion.

My work extends the traditional geographic theme of earth designed for humans, except that I examine a facsimile of this earth—a virtual geography that charts an array of representational spaces from the fantastical to the realistic, and extends along a continuum running from the individual to the corporate, from the contemporary individual "home alone" to the multinational corporation using the technology to advance its own ends. The invention of a virtual world represents human efforts to extend technical control not only over the social relations of others and of "nature" but also over the conceptions of space by which order is conferred onto what might otherwise seem an unimaginable void of meaninglessness. Humans have always been in place, yet they seek ways to extend themselves as part of producing meaning, as well as for the novelty of attaining different "points of view." This is an ancient and ongoing process, which I discuss more thoroughly in chapters 2 and 4. Contemporary electronically mediated communication, however, now increasingly substitutes for an actual physical going forth on our part. Communication technologies not only affect our experience of the world but also have concrete effects on our actions and the places of the earth.

Figure 1. Positioning the subject for a virtual future. Copyright Nippon Electric Corporation. Reproduced courtesy of NEC USA, Inc.

It is not my project here to examine how new forms of communication specifically change the planet's areal differentiation. I acknowledge the profundity and importance of these ongoing alterations, which are the subject of such works as W. J. Mitchell's *City of Bits* (1994). I am more concerned with users' *experience* of these concrete effects and changes. Such experience directly influences the meaning of community and politics and how individuals and collectivities imagine and make manifest the built world around them. *Community* and *communication* share related meanings, and both are evolving in meaning and conceptualization. The ability of VEs to destabilize identity formations has clear implications for what we mean by community, city, and public life. Communication takes place between or among people and people, or people and objects, but I am also using "communication" as something of a synecdochic model of the relationships among peoples and people and objects.

Virtual communications technologies exemplify a new form of "reaching out" capable of casting into question how we conceive ourselves to be in the world as *engaged* subjects. VEs simultaneously combine a sense of a confined place with a notion of a journey (Friedberg 1993, 29). Virtual technologies have the potential to remap or even collapse the modern experiential "distance" between subject and object, technology and social relations. Operating within a complex and disruptive period of globalization and "world economy," VEs further fragment the already shaky identities of peoples and places within the destabilizing "scenic domain" of the New World Order even as they help promote a renewed form of determinism and mystical thinking. These are expressed through "evolutionary" claims by which these technologies are positioned as autonomous from—and thereby superior to—the cultural "constraints" they are seen by promoters to supersede.

Virtual environments are a specific form of *interactive* communications environment, and they help effect a changing perception of our embodied relationship with the world. They participate in readjusting both the distribution of copresence, or face-to-face forms of being together, and more mediated forms of communication (Boden and Molotch 1995, 258). The VEs I examine most often are *immersive*—users don a head-mounted display (HMD) and may wear exoskeletal devices such as a vibro-tactile feedback glove or use a joystick. Users are also monitored by a position-tracking device that transmits information about the user's body position and motility to a computer.

Two possible objections should be addressed at this point. Understood as a confluence of social practices, optical technologies such as VR clearly have utilitarian applications; however, in no way do these applications preclude VR from being positioned within "the social imaginary" as a transcendence device. I am not suggesting that transcendence has a universal signifier. For some it is escape from the body, for others the planet, for others both. For some the route follows a path toward celestial or outer space; for others that "space" is ironically interior—whether the cyberspace on the "other side" of the computer's interface, or the "world" of the imagination, or both conjoined by the hybrid "space" of an immersive virtual environment. Second, though one might argue that VR is in its infancy—like the Ford Model T—the contemporary automobile is more like the Model T than dissimilar. At base the technology remains the same. At the Human Interface Technology Lab at the University of Washington–Seattle, scientists funded by commercial sources are working to replace cumbersome HMD technology with retinal scanning devices that would bathe the back of the eye with images. Such devices, which might look like a lightweight pair of glasses (which, of course, would still constitute a form of HMD), are in the early phases of research. Although the collective future of the various technologies that come together as a cluster in VR technology may be more sophisticated, its fundamental precepts and the theoretical and philosophical conceptions that underpin its developmental history will likely remain similar. Frank Biocca (1992b) notes that overall a "developmental logic" already circumscribes the several versions of VR under development. The "developmental logic" noted by Biocca that circumscribes VR initiatives directly connects to the notion of progress. VR already has a history, and hindsight offers a somewhat misleading vision of, or sense of progress about, the technology's increasing refinement and power. With most technology development, the early translation of specific research initiatives into technical form gives the outcomes of these initiatives their competitive edge. More people use them earlier, gaining familiarity with the technical forms. Early iterations of technologies consequently often seem "best" — to confirm a developmental logic — precisely because they have been deployed first as opposed to having been deployed because they were the best (MacKenzie 1996, 7). The technology's privileging of the logics of vision and sight is apparent. To wit, Biocca cites Ivan Sutherland: "The screen is the window through which one *sees* a virtual world. The challenge is to make that world *look* real, act real, sound real, feel real" (Sutherland 1965, 507; emphasis added). Biocca further notes that "the long-term developmental goal of the technology is nothing short of an attempt . . . [citing VR scientist Fred Brooks] to fool *eye* and mind into *seeing* . . . worlds that are not and never can be" (emphasis added). "An array of light on a visual display becomes a lush landscape in the mind of the *viewer*" (Biocca 1992b, 27; emphasis added). As the various emphases indicate, vision is and will remain a fundamental, even primary, precept in virtual worlds, no matter the level of sophistication attained.

I do not ignore sound and touch in the chapters that follow, but I would suggest that they too are iconized by optical technologies — as much as possible made over to support long-standing implicit Western associations between knowledge and certain forms of seeing that lead to the assertion "seeing is believing." In the video game *Maelstrom*, for example, when the user manipulating the "good spaceship" fails to rescue the supply ship from destruction, the program always growls, "you iiidddiotttt!" When I download E-mail from my on-line service provider, a cheerful and lilting woman's voice always announces, "mail waiting!" This is the nature of much programming; there is little variation save that within a predetermined range. Sounds in a VE are related to the spatial relationship between user and icons. They are always the same for any one situation. When the cupboard door in the virtual kitchen is opened, it always goes "whoosh." When closed, it goes "thunk." In this, digital sounds in a VE operate as "aural icons." Moreover, as audio theorist Steve Jones argues, the very jargon of VR excludes the aural (1993, 239), and the creation of VR "can be understood as part of the ongoing technological visualization and deauralization of space" (246).

Touch or tactility in a VE remains a very visible tactility. One orients oneself visually, and as with sound, touch is made a proof of what has first been seen. Margaret Minsky (1984; Minsky et al. 1990) of MIT has pioneered research on touch and gesture in VEs. Minsky's system uses 2-D gestural data to classify gestures according to intent — whether they select, move, or indicate things or directions. Again, this is a very iconic or visual understanding of a nonvisual sense. She is also interested in "force feedback" — how virtual objects can be given the semblance of a resistant materiality. At the University of North Carolina at Chapel Hill, an application under development by the Department of Computer Science uses the PHANToM force feedback device. The device requires donning a special pair of glasses. When I held the feedback joystick device, in a demonstration of the technology in early 1998, I experienced the computer pushing back as I navigated across a virtual terrain with the help of the glasses. Going up a hill, for example, causes the hand unit to push back more forcefully than going down. The steeper the hill, the greater the force of the push back. In a sense, physical exertion gets translated into a rheostatic experience.

All branches of VR development — be they entertainment, scientific, industrial, medical, or military — currently employ some kind of HMD as part of their research strategy. Suggestions that VR's real promise is a corroboration among the senses fail to consider the disjuncture between subordination and corroboration. Subordination to the visual really points to the coordination (and domination) by the visual of our other bodily faculties and senses. VR privileges sight, and other senses play a subordinate role to it. It is misleading to imply that a "haptic" VR, integrating the senses, is somehow imminent. It is indeed the "promise" of VR, but a central premise of my work is that the ways in which this promise is hyped mislead many observers into surmising that "promised" advances—which are the stock and trade of science fiction—are just around the corner, or already available.[4]

Stage Set

In May 1995, at the "first ever" "Futures Forum" convened to discuss "the future of virtual reality conceptually and technologically" (Mecklermedia 1995, 14), Dan Duncan, a consultant and muse to industry players, described VR as "the emperor's new technology." He noted the now familiar high level of media attention mixed with hype accorded VR, but echoing critiques of technological rationality, he suggested that no one really knows what VR is, those responsible for its development included. Although "interactivity" remains a pet industry buzzword to describe VR's effect, originally, Duncan continued, this term had referred to people interacting with one another through VR as a medium. However, Duncan suggested with regret that "the VR community," faced with tough bottom-line decisions imposed by a skeptical marketplace, has redefined interactivity to refer to people-machine interactions—though one could as easily argue that this is just what many users wish: to interact with a simulated world subject to their own willful control. In an effort to direct the forum's collective imagination away from virtual scenarios overly and overtly influenced by military applications, Duncan invoked VR's utopian promise to "put words into a different medium" as part of a realizing of our "human birthright" of "an invisible world we have a chance to bump up against in the dark" (my transcription).

The same forum at which Duncan spoke afforded me the opportunity to experience the technology firsthand. Of the fifteen virtual environments I sampled, the most realistic was Division Incorporated's military simulation, "Weapons Systems in Virtual Trials," developed in cooperation with the U.S. Army Research Laboratory (ARL). According to Division's (1994) product monograph and application brief,[5] ARL is using VR to develop a system for testing, evaluating, and refining new designs for weapons and equipment before constructing actual prototypes. When the system is fully developed, infantry teams will enter the VE to fight simulated battles using as-of-yet unbuilt technologies. "An infantry team assigned to a new anti-tank weapon, for example, will be able to square off against a tank crew operating tomorrow's version of the M1 Abram tank. The lessons learned will be applied to improve both equipment designs and operating procedures." As the two-dimensional stills of figure 2 indicate, the HMD-wearing user is immersed in a VE incorporating both urban and rural environments that are populated with enemy

Figure 2. Weapons systems in virtual trials. Copyright 1994, Division, Inc. Images courtesy of Division, Inc.

tanks and paratroopers. Armed with (virtual) antitank and semiautomatic weapons controlled by a handheld device, the user-soldier responds to enemy attacks as she or he navigates the representational space. High-quality three-dimensional sound helps heighten the sought-after sense of realism (ibid.).

I treat VR as a set of technical developments and social practices that organize my examination of philosophical and social issues pertaining to communication, information, language, space, and vision. My own immersion in this VE was not intended to verify the quite plausible corporate claims made for it as a military theater for testing weapons systems; I sought only to experience for myself the degree of realism attained. I use a military application to exemplify the technical features of the technology, and I do so here *because it is not a commercial application.* It is more sophisticated in part because of the U.S. Department of Defense's deep pockets — and therefore the technological cutting edge available to ARL researchers in creating their virtual worlds. Costly refinements yield complex, highly realistic virtual worlds, whose design offers an excellent set of examples for probing the technology's built-in philosophical assumptions. More money makes more interesting environments to theorize.

In Ralph Schroeder's *Possible Worlds: The Social Dynamic of Virtual Reality Technology* (1996), the author points out that competitive research strategies, which transcend nations and distinctions among the academy, the military, and private enterprise, are what drive VR development. Such strategies *synthesize from the outset* hard science research and commercial applications in the belief that VR will transform information technologies and communications networks. Schroeder looks only at commercial applications, for he argues that the costly sophistications built in to military VR largely preclude easy adaptation into commercial products. Yet Schroeder places considerable emphasis on games, for they drive the commercial VR industry. I find it more than coincidental that so many current VR games replicate the thrust of the military training application I describe here — find the enemy and kill "him."[6] I concur with Lev Manovich (1995a), who notes that all the key elements of modern human-computer interface devices have been developed by the military. Their history has less to do with public entertainment than with military surveillance and the increasing use of technology as part of a quest for control. In turn, the mathematical theories and methods of engineering that underpin systems analysis generate *closed* models of the world. Interlocking systems subject to mathematical analysis, together with mathematical models on which computation is based, had strong appeal to American military elites during the 1950s and 1960s. "Global politics became a system that could be understood and manipulated by methods modelled on—or at least justified in the language of—systems engineering. The computer provided a core around which a closed-world discourse could crystallize" (Edwards 1989, 140).

The Army Research Lab's application is a "mix and match" of cultural and philosophical borrowings, and it showcases the traits of VEs that I examine hereafter. These traits draw together specific concepts of space, place, and landscape. The technology suggests a world composed of light and almost entirely reliant on vision, yet one in which no human bodies or other living organisms are present.

Because the well-funded ARL is the client, this VE's sense of photorealism is very high, though there is a lingering sense of being inside an illuminated cartoon. After adjusting to the sensation of wearing the head-mounted display, and using the handheld joystick with which one moves one's virtual position "forward," "backward," and "up" and "down" in "space," I began to look around and take stock of "where" I was. I "entered" the VE at a point halfway down a building-lined street not far from what appeared to be the main intersection of a small town. This town seemed to be in the middle of a rural environment whose nearby topography was flat, with undulating hills in the distance. Fencerows and stone walls lined the highway that led out of town. In navigating my way around, it was possible to "fly" along the street, either in the direction of the main intersection, or, by turning my body around in real space, in the opposite direction toward the fields on the outskirts of town. I could adjust the height at which I flew, either to conform to the distance between my eyes and the ground I normally experience, or higher or lower. The sense of flying was unlike walking, but an initial sense of strangeness was mitigated when I adjusted the virtual height between my point of view and the ground to mimic that of real life. In a sense, because the VE is quite successfully programmed and designed to experientially and emotionally engage the user, it affords a "real" experience, yet also a closed one subject to manipulation and control.

Colors in a VE are exceedingly radiant—this is a world composed of light. In the military simulation I am describing, buildings on either side of the street were highly textured. This offered a sense of realism. However, it also deflected attention from the two-dimensionality of the images on display. Minus their bright textures, these images would have appeared more stark and less conducive to stimulating my participation — hence the emphasis on architectural details incorporating a high degree of geometricity. The inclusion of multicolored brickwork, detailed fenestration, balustrades, porticoes, fence rails, camouflaging paint schemes on aircraft, and the like seems a necessary component in achieving the user's willing suspension of disbelief that the VE might constitute a "real world." Objects from the natural world, with their greater fluidity of form, are more difficult for the technology to represent in a realistic fashion. Accordingly, they tend to be fewer in number than representations of manufactured items and are often positioned to be seen at a sufficient distance so that the eye accepts their representation in fairly short order. In the ARL simulation, trees were positioned in the middle of a distant field one could not cross. The computer program did not allow me to approach a tree to examine the quality of the representation "close-up."

As I was examining the street facade of the village (which appeared abandoned by "local" residents), an enemy soldier materialized in front of me. "He" drew his rifle in anticipation of firing. This soldier-character functioned somewhat like a target in an artillery range, and it was possible to aim my virtual gun and annihilate the image. Had I waited too long, however, he would have fired on me (or, more precisely, in the direction of the virtual coordinates "where" I appeared to be positioned within the VE), and the computer would have registered my presence as "terminated." In trying out the ARL application, I found it impossible to ignore the lack of women. Men design a killing field and testing ground for other men. Once "inside" the simulation — a world clearly gendered as male — it is images of men and machines with which a presumably largely male user group interacts and "terminates."

Proceeding through the intersection toward the other side of town, I was able to enter the console area of a virtual tank and operate its controls to shoot down enemy helicopters. The explosion and crash of these machines, though sounding the same each time, was quite realistic, both spatially and in time. One could say that a partial sense of place is achieved in this VE. I fairly quickly suspended my sense of disbelief that I was walking in an image, even as I remained aware that this was so. As discussed hereafter, this suspension of disbelief may relate to the technology's newness, and an interest, even eagerness (particularly on the part of individuals operating within a highly mediated and technological culture) to interact with this latest technology and master it at some level. Certainly, if landscape is understood as the visual aspects of a place, then a strong sense of landscape is achieved, albeit one that is highly geometric in execution. However, this visual sense of looking into a scene constructed according to laws of geometry and perspective is not the same as how we see the real world. A central difference is the intense luminosity of a VE. The brightness is dazzling, a subject I will return to in chapter 5, and my prolonged experience with the technology over a period of three days left me with severe headaches that lasted for hours at a time. For several nights after my three-day stint in VEs, while lying in bed, when I closed my eyes and "envisioned" a scene, I was able to zoom it forward and backward at will, replicating in my imagination what VR appears to make possible.[7]

A more substantive critique flows from my self-questioning as to "where" I was when in this VE. Yes, I was conscious of interacting with a simulation whose design cannibalized features of built forms so as to offer something familiar to users. I did not think the town was located in Africa or Asia, though I noted that enemy craft appeared to bear Arabic markings. Rather, the town might have been an ersatz English village—complete with its turreted stone clock tower and recessed Regency-style paneled doorways of houses giving directly onto public sidewalks without the intermediary (North American) spaces and features of front porches, lawns, or fences. This English village had been relocated into an environment that seemed like an arid plain in the American Southwest, a juxtaposition that prompted me to recall the importance of the relationships between value and real places, and the cultural and historical contexts that lead to the making of the latter. Moreover, because my body was not present in this VE, the experience seemed to matter less than real life—for example, it did not really matter much if the soldier shot me or not—even though I was aware that the technology promised a control over experience not available in real life. Yet this form of control is precisely the technology's value for the Army Research Lab, which hopes to test new technologies at minimum cost to the health of personnel and to the bottom line. Experiences within this VE can be monitored, recorded, reordered, and replayed in a way unavailable to embodied experience and the "messy" world of contingency and "natural" intrusion.

The greater the number of pixels per inch, the finer the grain of the image, the closer VR comes to a correspondence with "the real." Yet the meeting of the English village with the middle of nowhere begs the question: correspondence with what, with where? Correspondence in VR need not be with any real place on the earth, but rather with imaginary places and circumstances made to seem real enough by an appeal to aspects of visual perception responding to texture gradients of surfaces and so forth. In VR, articulation or disarticulation with "the real" is a part of a program reflecting the relative power of clients and designers to ignore at will those aspects, contingencies, or contexts of, for example, English villages and American Southwest deserts that are inconvenient, undesirable, or incompatible with instrumental goals.

In the chapters to follow, an argument will be developed that virtual technologies manifest postmodern sensibilities, ones that are also reflected in academic theories that, in the words of Lorenzo Simpson, profess "an allegiance to the idea that all of reality is a social and linguistic construction" (1995, 142). If one's sensibility were to correspond with Fredric Jameson's findings about postmodernity (1984) — in which a postmodern period or sensibility has emerged following and because of the completion of that part of the modern project progressively committed to subduing the natural world — then such a sensibility would no longer value what is old, including the natural part of our lived worlds. Jameson's understanding of such a contemporary cultural sensibility can be applied to Simpson's assertion that postmodern desire "is the demiurgic desire to be the origin of the 'real' " (1995, 140). Postmodernity collapses binary oppositions such as illusion and truth, appearance and reality, culture and nature. Now, if nature is believed to no longer exist in any meaningful way, or is "written off" in part by arguments that it is only a cultural construction, then a postmodern sensibility desiring to be the origin of the real would have to substitute something synthetic for something real in its production of reality. Hence, virtual or artificial reality and its accompanying "fake space,"[8] suggesting a conscious awareness that humans have (or wish to) become the "authors" of their own ontological ground. Such arguments receive support from the ongoing transformation of nature into commodified resources, and VR the technology instantiates postmodern and poststructural theories that insist there is no world beyond the text.

At the same time, an increasingly widespread difficulty with metanarratives heightens insecurity, having the effect of tilting cultural inclinations ever more strongly toward technical fixes for this insecurity (Simpson 1995, 140). Enter VR as the technico-cultural fix invented by a postmodern sensibility both as a bulwark against uncertainty instigated by the perceived death of the real and as an uncanny artifact created by a latter-day nostalgic Dr. Frankenstein in search of a means of producing a seemingly vanquished (meaningful) reality. In all of this, VR suggests that a premodern realism can be recuperated through using modern conventions of representation.

A Machine for Performance

Users of nonimmersive telecommunications, such as IRC and other chat room environments accessed via Internet connections, experience a "feeling of 'place,'" that they are "in" something, are some "where" (Dery 1993b, 565). However, once accessed by interface devices such as computer screens or head-mounted displays, if there is no material "where" within cyberspace other than the debatable materiality of electronic data constituted in light, electricity, and zeros and ones, is there a "geography of cyberspace" worth examining? A strict empiricist might limit investigation to the mappable — to the visible spatial reallocations of people and things to which the technology might contribute. It is the combination of cyberspace's initial invisibility and its power that generates a desire to make it visible. With VEs, cyberspace appears visible and seems to make sense to positivist approaches, its representationality and metaphoric qualities notwithstanding.

One of the prime engines driving the development of virtual technologies has been the relentless demand for an ever more efficient way to move capital within a global economy. While it is no doubt "efficient" for capital, reformulated as infinitely *flexible* data, to move at the speed of light[9] across a variety of geographic scales, our embodied reality does not respond in as salutary a fashion to such ephemerality. To achieve this degree of flexibility, both capital itself and the spaces within which it is represented have taken on a much more fluid identity. Derivatives are much more ephemeral than gold bullion, and dataspace much more so than a vault in a bank in which something even so marginally tangible as banknotes or specie might be stored.

The mutable identity so necessary for competitive advantage and rationality under advanced capitalism is exactly *not* what people need from the places of their lived world. Robert Romanyshyn makes the exceedingly practical observation that in everyday living, "we count on things to keep their place," and that we are able to do so because "we have lived our lives with them in this fashion . . . in their fidelity to us they function as extensions of ourselves" (1989, 193). Things remaining in place help root our sanity. In VEs, however, depending on the power of the software, what appears to me as a chair may be for you an animal or perhaps not even present at all. Not only need things not keep their place and form but neither, it is proposed, need people. I may appear to you as a funnel cloud if angry or a happy face if sad. Indeed, development is in progress that would allow users to represent themselves as multiple identities simultaneously; hence I might appear as a happy face to you and a funnel cloud to her. What is lacking in arguments advanced by those promoting this profoundly individuated virtual "freedom" and "pleasure" to play with identities, subjectivity, and geographies[10] is a sustained consideration of the meaning and context of self-control of our actions, along with any sustained interrogation of the consequences for social relations beyond the scale of individual access. If, as users, we are truly to be so fragmented within the "play" of VEs, then *which* aspect of our identity is it that will morally inform our actions so that we do not inadvertently hurt or damage those people and things we care for in this unbridled free rein of identity? But then, with fractured and multiple identities, precisely whom or what are we speaking about? Places of the body are contingent things. Places and people may change, but this all takes time, a key lesson dispensed with in cyberspace. It is true that in the fragmentation that constitutes a significant portion of everyday contemporary Western life, most people are forced to juggle several roles during the course of their daily affairs. They are mothers, workers, lovers, consumers, holiday makers, queers, executives, religious fundamentalists, part-time teleworkers. But I suspect that for most people, a central core of self remains to do, or at least to *authorize*, this juggling of identity components. This "core" is located within the modern understanding that one is "present to oneself"—imaginatively and physically. The embodied and imaginative Hobbesian "Author"—a naming of that which ontologically may precede naming—still watches over the "Actor" on the myriad stages of life.[11] Although such a central basis of self may also remain "behind the scenes" within VEs to determine which identity is to be donned or acted, this recognition on its own begs the question of bodily truth, as witnessed by others with whom we each must deal (see figure 6, in chapter 4). Whether holiday makers or mothers, our bodies remain with us both as testimony to who we are and as a unifying dimension of ourselves within social polyvalency. Not so in VEs, where users' bodies, if represented, are only components of simulated digital space and need not be tied to any representational public facade the self may employ.

To the degree that a VE-as-text remains unable to confirm the centrality of users' bodies, it sets the stage for a disempowering relativism (see

Jackson 1996). VR's potential for allowing experiences of extreme poly-valency and polyvocality—of spaces represented, identities "performed," and multiple outcomes rearranged and replayed at will—is presented by academics and promoters alike as ushering in a carnivalesque world of pleasure and play. Don Mitchell notes that the idea of culture operates "to control and order aspects of an unruly (but nonetheless highly struc-tured) world" (1995, 113). As a cultural technology, VR not only allows a new range of identity performances but also functions as a technology of social control precisely through its promise of polyvocal polyvalency. Assuming the price of admission, each user may play with identity/space. She or he may do so, however, only when subject to a naturalization, and hence a forgetting, of the physical laws governing the technology as well as the social relations organizing its allocation and the ideation of its programming.

An increase in spatial segmentation—and emphasis on distinguish-ing between private and public, actors and audiences, self and "external" world—reflects a deepening sense of self (Tuan 1982, 9). This deepening resonates with Beck's (1992) argument that modernization progressively increases individual articulation. Tuan finds that though the self is an entity, it can be segmented without end. Although I have stated that the "Author" (or modern self, or "I") remains important—despite the frag-menting impact on identity formation such spatial segmentation may effect—a principal result, even goal, of such segmentation within virtual worlds involves divorcing the sense of self from one's own flesh. The use of the term "self" is historically conditioned (Taylor 1989, 32). Applied to the modern individual—a result of the development of a certain phase of scientific, political, and economic thought (Williams 1983, 164)—the self has come to mean a being "of the requisite depth and complexity to have an identity." The self can never be completely articulated, in part because one is never a self on one's own; however, one's self is consti-tuted by and within the language community of which one is a part (Tay-lor 1989, 34). If this language community increasingly is to be experienced within on-line immersive and nonimmersive optical and text-based en-vironments, and if the meaning of the self is never entirely articulated, then it is at least arguable that a sense of the self may in part be gained from the use of such electronically mediated technologies. It would fol-low that the means or criteria by which we distinguish ourselves from machines are themselves in the process of being redefined via a long-term and increasing reliance on virtual communication technologies (Turkle 1995). Even as we experience increasing spatial segmentation among our human selves, the boundaries between the self and the technologies it uses to transcend this segmentation seem to begin to blur.

The Ethics of Virtuality

I am interested in the ethics of virtuality. VEs do not only substitute, represent, or simulate the concrete and fantasy places within which the embodied subject participates in the lived world. VEs also represent an alienation of political and ethical values and meaningful practices to the degree that these values and practices are reformulated as technology's ends (Simpson 1995, 164). For example, freedom is highly valued, but rather than creatively engaging with the contingent limits to freedom, VEs propose we surround ourselves with freedom as a commodity we produce *as if* gods. The world's material forms seem to matter less in VEs, reflecting the long trajectory within the dynamic of modernization to free social actors from structures (Lash and Wynne 1992, 2). What meaning might such a technically reengineered freedom have for the modern subject, in what may amount to on-line schizophrenic identities-as-transient-performances, or the illusion of freedom from structures ironically achieved through science and its application to a technology "wherein" one's body can seem a barrier to freedom? Even given the "pursuit of happiness," pleasure can be distinguished from happiness or freedom. Pleasure, pace Bentham, need not be held as the highest goal, the more so if the means by which this individuated pleasure is gained remain unexamined, or accepted without considering what is forgotten or yielded in exchange.

I introduce ethics with some trepidation, aware that a culture driven by access to pleasure-as-consumption unceasingly redefines this area of concern variously as reactionary, Luddite, or totalizing. Yet I want to resist the trend not to question technology, whose increasing power makes critique difficult to the point that those who consider its agency may be labeled "eccentric" or "dystopic." There remains considerable uneasiness, both within the academy and elsewhere, about considering links between ideology and technical practice (though see Feenberg 1991). Perhaps this is because many see no alternative other than an ever increasing dose of technology, given the death of metaphysics and retreat of meta- or "master narratives" under an apparently benign scientism. Some prefer not to bite the hand that feeds. This kind of fatalism flows from a philosophical *heteronomy*—the condition of being under the rule or domination of another (Curry 1995). On the one hand, scientists and other users of technology operate within a Utilitarian value system in which the successful workings of technology outweigh the value of its critique. On the other hand, individuals are asserted to be autonomous, moral beings. Following Immanuel Kant, agents *ought* to "act out of the belief that an action is a good action" (74). Such autonomy places the agent outside of the system, which is justified on the basis of its ability to "deliver the goods." Ends justify means, and partially as a result of the quasi-religious status of science, a scientism fosters acceptance of according autonomy to systems or technologies. Application of theories such as cost-benefit analysis and environmental or social impact assessments permits scientists to assume the sovereignty of their own positions, yet to deny their own hand in first making and then reifying this assumption. The consequences of science and military and industrial technologies such as VR, however, now lead to an unprecedented set of hazards and risks (Beck 1992). Inquiring after the ethics of virtuality, therefore, is part of a larger critical project required for fostering and foregrounding reflexive arguments that might help resist the often unquestioned trend to substitute parts of the lived world with technology.

Considering how the lived world is spatially ordered requires taking account of the philosophies, belief systems, ideologies, and discourses influencing the permissions, impositions, and negotiations that result in the spatial demarcations with which we live. The spatial dimension of communications technologies is central to these dynamics. The physical and social demarcations that result from these permissions and negotiations reflect specific conceptions of space — ones designed to impose social order and confer identity and meaning on ourselves, the world around us, and the larger cosmos within which we "float." A central concern of my research is the distinction I identify between communication and existence. Jean-François Lyotard argues that a high price has been paid in episodes of terror flowing from the West's belief that "the concept and the sensible . . . the transparent and the communicable experience" might be joined into one (1984, 81–82). What he points to is not that "idea" or "concept" should not inform the body or sensation, or that art must not inform life, or literature advance conversation. Rather, Lyotard grasps the need to *remember* that idea and matter stand in relationships to one another in which there are no guarantees. As Fred and Merrelyn Emery (1976, vi) point out, communication is a secondary property. Although communication is a necessary condition for people to act socially, on its own, communication can never be a sufficient guarantee that this activity will occur. We do, however, bring to our reading and writing practices an always already embedded "preunderstanding that mediates between the knower and the known" (Chang 1996, x). VR, with its promise of an ersatz world in which any guarantees are based only on the virtual environment's programming, is centrally implicated in these issues.

Organization

This work is a form of criticism that generates theory. A pivotal connection exists among criticism, history, and context, and in many places, I incorporate necessarily selective historical accounts to ground the critical theory being proposed. Chapters provide progressively widening inquiries into the significance of emerging virtual technologies and VEs, and the challenges they pose to existing concepts of identity, subjectivity, and space, and to actual human bodies and places. Because VR, in part, organizes my inquiry, to familiarize readers chapter 1 provides a critical history of the technology and its development within a largely American context. Having offered readers a sense of the twentieth-century material and ideological conditions underlying VR, I move in subsequent chapters to broaden the range of ontological and epistemological inquiry into the *idea* of VR, and to shed light on the current cultural desire for virtual worlds.

I italicize the word "idea" to call attention to the philosophical underpinnings and discursive strategies represented in VEs, for this work is not an inquiry into cognition. I avoid the kind of causal arguments implied by the concept of cognition. I do believe that a full study of perception as it materially and conceptually relates to the user's situatedness and embeddedness within VEs is a project very much worth undertaking. It is, however, beyond the scope of this book, which is more concerned with making connections among the idea of VR, real places, human bodies and ethical actions or the lack thereof.

Although I am interested in theorizing relationships between communications and how reality is culturally constructed, I acknowledge communications as having often been about the waging of wars (Mattelart 1994) and the achievement of strategic advantage over military, economic, social, and political enemies. VR technologies are often offspring of the U.S. military and its ongoing interests in power and control. Commercial and consumer spin-offs retain certain design aspects predicated on their original military applications, though VEs are beginning to extend the spectrum of what can be transmitted by widening the range of representations that communications devices are able to convey. In their role as an information technology, VEs will broaden the bandwidth array of sensory information users may transmit about themselves as they begin to extend their selves conceptually via these image technologies across a global terrain. Despite their military genesis, VEs are inflected not only by a technological imperative, or will to power via representational means, but also by literary influences and countercultural utopian aspirations. The closing of the western frontier, the development of flight simulation and digital computation, the role of science fiction, and that of psyche-delic drugs as access mechanisms to a transcendental state have influenced the development of contemporary VR. Chapter 1 examines these influences and traces significant social processes and actors leading up to the current state of affairs in the "virtual world."

Widening the scope of consideration of influences on VR, chapter 2 focuses on late-medieval, Renaissance, and Enlightenment cultural and optical technologies that in various ways foreshadow aspects of today's virtual worlds. The focus is selective. Specifically, I examine the camera obscura, the magic lantern, the panorama, and the stereoscope, and I look at ways that VR mixes and matches aspects of these precursive tech-nologies, which have at different times been understood variously as confirming subjective interiority as a center of truth, or as establishing the "correctness" of an absolute, exterior, and divine source of illumina-tion. Even more than TV, from which VR borrows, for example, CRT tech-nology, VR can be understood, in part, as a constellation or convergence of several earlier technologies as well as some things new. The case of cin-ema is instructive. By the 1890s, audiences had developed an appetite for all manner of optical entertainments (Manovich 1995a). These in-cluded magic lantern shows, panoramas, dioramas, stereoscopic dis-plays, and a range of other devices less well remembered today such as the thaumatrope (literally, magical turning), phenakistoscope (deceitful view), praxinoscope, zoopraxiscope, and so forth. Yet what Manovich terms the "dynamic screen" of cinema, TV, and video—the display in a rectangular screen of an image that changes over time—had been sug-gested in earlier devices but not fully present until the introduction of cinema technology. The dynamic screen also informs VR. However, the sense of a window onto another space—for example, televisual space—is collapsed into the VR user's point of view. Cinema and television's dynamic screen requires identification of the viewer with a screen image (ibid.). As with the panorama, however, in VR, one becomes part of the VE, and the identification with a screen character is not central to the experience. It is also possible in cinema and TV to identify with the cam-era's positionality, and hence the cultural purchase suggested by the phrase

"I am a camera." Although an immersive virtual experience might be theorized similarly—one actually wears a camera—the salience of this experience requires setting aside any such belief and imaginatively extending oneself into the world of objects, shapes, and representations of others in at least partial recognition that one is also represented by an icon to other users. Although cinema and TV reify an already existing belief that seeing is believing, the just-noted cultural and technical disjunctures and connections suggest looking to earlier technologies for the philosophies and aspirations contained within their designs, which continue to inform new optical technologies—VR and TV included—though in different, and at times contradictory, ways.

Because I am looking at how earlier technologies were positioned to mean different things at different times for different peoples, the opening section of chapter 2 examines technological determinist and social constructionist arguments as straw men for one another and suggests ways in which the agency of technology might more productively be argued. I retain a skepticism about the erosion of conceptual boundaries between humans and machines. Nevertheless, though technologies are social constructions, they are more than only this, and we commit a disservice to their understanding in subsuming them under social relations. The issue of technological determinism is related to the modern project's success in divorcing science and technology from politics and social relations. Therefore, making use of Ulrich Beck's (1992) analysis of risk and modernization, Bruno Latour's (1993) analytic history of early-modern efforts to parse technology from politics, and John Searle's (1995) distinctions between natural or "brute" and socially constructed facts, I examine how VR, as one of the modern hybrids Latour identifies, draws together technology, politics, and social relations. Continued belief in the myth of a value-free technology, however, propels "risky" utopian notions that technology might constitute a kind of "natural" home apart from the embedded social contexts of politicized social relations. VR is a most American technology, operating as it does to contemporize and extend the notion of a sense of individual renewal coupled to encountering a spatial frontier, and the final section of chapter 2 updates the notion of an American insistence on an "electric sublime," first proposed by James Carey and John Quirk in 1970 as part of their exposition of the ongoing utopian desire for control over nature.

Chapters 2 and 3 are companion pieces. Building on the work of the previous chapter, chapter 3 sets out three underlying considerations that inform this book's review of optical communications technologies. I first consider the meanings of ritual and of transmission as these have been applied to communication and its technologies. Second, I consider the relationships among conception, perception, and sensation, for VEs have the ability to collapse or rework distinctions among the three — implicating a conflation between the conceptions of others and the perceptions of the subject user. VR is promoted by industry players as a representational space, and the technology, in part, depends on naturalized conceptions of space, place, and landscape for its cultural saliency and reception. Which concepts inform VR, and how they do so, are important, the more so given the imprecision and nuance of meaning attending these different concepts. Therefore, I discuss different concepts of space, consider the distinctions and similarities among space, place, and landscape, and suggest their importance to VR.

Because immersive virtual technologies are a visual representation of spatial reality, chapter 4 inquires into vision and sight. Virtual environments are the visual worlds that VR users encounter. Because virtual environments are so visual, I organize this chapter as a discursive history of the roles of vision and sight in Western understanding, and how vision and sight inflect Western concepts of space. I examine the power that vision is accorded both epistemologically and as a metaphor. As a material technology relying on visual surveillance mechanisms and activities, Jeremy Bentham's eighteenth-century panopticon and the self-disciplining effects it partially produced and induced are worthwhile to consider in relationship to how such effects are extended within contemporary VR, and the surveillance and transcription possibilities it supports. Users seeking pleasure by recourse to virtual worlds consent to engage with the technology at a bodily level, and more so than in the case of watching a film or TV or listening to the radio. Users are complicit in, and yield to, their disciplining by the machine. VR thereby forces a reconsideration of the relationship between pleasure and surveillance. Within virtual environments, pleasure and surveillance are in an as yet underacknowledged dialectical, and not oppositional, relationship.

Space is often conceived in visual terms, and the discussion of absolute, relational, and relative conceptions of space opened in chapter 3 continues in chapter 4, with a history of the various strands of classical Greek thought synthesized in the Euclidean geometry that forms the basis of absolute space. I consider how these conceptions are built in to virtual technologies, and how a residual dichotomy of visibility and invisibility enters into their production. The competing theories of sight, vision, and sensation advanced by Bishop George Berkeley and James J. Gibson anchor an extended discussion of the roles of human embodiment, motility, social relations, modern formation of identity, space, distance, and theories of hardwired perceptivity in VEs.

Chapter 5 extends a discussion begun in the previous chapter of the pros and cons of metaphor—which is central to the production and meaning of social discourse. Although widely thought of as vision and image technologies, virtual technologies depend on a base of software and code. VEs exemplify the attempt to merge language and vision as representations. I include a history of metaphors of light that have helped shape an eventual acceptance of "virtual living," and a shift from subjectivity to image. Classical assumptions that variously position the viewer *in* the light or looking *into* the light, and modern ones of being *in and of* the light, are considered for ways in which they inform the spatial understandings built in to virtual technologies.

With chapter 5, however, I also begin to return attention toward considering current links between VR and other cultural and political phenomena. Mikhail Bakhtin idealizes the medieval carnival and its embodied laughter. In doing so, he authorizes a questionable homology between marketplace carnivals and the print technology of the modern novel. I examine this homology as an academic instance that sheds light on a parallel effort, exemplified by VEs, to relocate what is meaningful in real places to a textual and representational subset of reality. I critique Bakhtin's work by proposing a disjuncture between a carnivalesque embodied reality and the subjective virtual experiences VR provides, and I argue that a belief that language constitutes "the real" is uncannily similar to the technicized truth claims made for VR by its promoters.

Chapters 4 and 5 discuss relationships among space, vision, sight, language, and metaphor. These relationships are intimately connected and difficult to pull apart. When, individually or in combinations, they are mediated by electronic technologies, multiple overlapping effects are produced that are complex to deal with analytically (Pickles 1995, 3). This difficulty suggests the value in organizing these two chapters as different "slices" through the same phenomena, in order to better grasp articulations or the lack thereof among them. Chapter 6, organized around identity and embodiment, their relationship to place, and VR's possible influences on them, offers a third, complementary approach to the issues at hand. I examine the place and absence of human bodies in VR. Without bodies we would have no place to take. Our bodies—which at least are coterminous with, if not preceding, the existence of language and language as a technology—situate a way to understand how history in general can be used by the powerful to oppress others and reify dominant discourses. Yet as a technology and a social practice, VR exemplifies a Western yearning for transcendence via achieving physical and cultural imaginative remove of the subject's mind from her or his body. A culture that increasingly accepts the mediated imprint of technical rationality also convinces itself that its peoples' various bodies are almost entirely the product of social relations and thereby only texts. Ironically related suggestions that VEs will offer a transcendent sphere of escape from the "here and now" are held against the meaning of transcendence as coming out of engagement with the real world. I theorize the light-filled world of a VE as updating the classical spatial positioning of truth seekers who are dazzled by transcendent light. VEs suggest that direct access to knowledge reduced to information is possible by visible means alone, yet human bodies anchor our ability to extend ourselves imaginatively into the world, and to do so in an ethical manner. This understanding is held against a competing assertion by proponents of virtual worlds that extensibility itself and the sphere of communications are all that really "matter." This assertion, inadvertently perhaps, speaks to the intersection of transcendence and capitalism, and VR as a transcendence machine supports thinking about our bodies as somehow in the way of a capitalist future discursively positioned as one of globalized "flows" of information and data.

Schizophrenia reflects a belief that the mind can control the world around it, indeed that the distinction between mind and material world does not exist. The continuous circulation of information supported by ITs and VEs, and the development of intelligent agents and other networked software devices that will operate within them, coupled with VR's ability to operate as a potentially infinite model of fragmented identities, contributes to the legitimation of schizophrenia as an acceptable model of social relations. VEs also suggest that psychasthenia — related to schizophrenia, and an experiential merger of the place of one's body with the wider lived world — might be an acceptable modus vivendi for contemporary (post)subjects reorganizing as images. Finally, I probe the political implications of the contemporary rush to virtual living. Much of VR's cultural appeal lies in its placeless utopian promise and premise, which I criticize through a review of earlier utopian schemes. Because the technology can operate to disarticulate identity from self-embodiment, traditional theories of resistance — implicitly according a central place to the role of human bodies — need to be rethought in face of the current drift toward all things virtual.

1. A Critical History of Virtual Reality

Do cyberspace and VR have a moment of invention? Are they a decisive break that sets them apart from telephony, TV, and digital electronic and communications technologies from which they are partly cobbled, imagined, and extended? Where might an account of the cultural trajectory informing the electromechanics of VR arbitrarily begin, given that much of the "buzz" surrounding it is concerned with asserting its novelty, thereby to author and secure its future, rather than to acknowledge a past? The 1990s' surge of interest in the phenomenon of cyberspace is heightened by promoters describing it as a new frontier, one open to exploration as well as colonization. Within the academy, and often just barely removed from the commercial hype, cyberspace has been conceived as "a globally networked, computer-sustained, computer-accessed, multidimensional, artificial or 'virtual' reality" (Benedikt 1992b, 122). An increasing variety of virtual technologies offer windows onto these cyberspatial environments, defined by Frank Biocca (1992a, 6) as ones in which the user feels *present*, yet where things have no physical form and are composed of electronic data bits and particles of light. Biocca also suggests that VR "can be thought of as a goal in the evolution of communication and computer technologies" (6). To date, no single technology, machine, or social practice circumscribes the assemblage of emerging immersive VEs, yet much interesting writing about cyberspace and VR assumes the technology as a given.

This chapter examines the human agency that makes this technology possible. It is a narrative informed by three assumptions: first, that the technology represents an instance of an ongoing (Western) motivation to alter conceptions of space; second, that its development is inflected by a desire on the part of a disembodied, alienated subjectivity for transcendence from bodily limits; and third, that this cybernetically achieved transcendence — as reflected in the 1980s cyberpunk desire to leave the body, or "meat," behind and float as pure data in cyberspace — is also a vehicle for merging a hyper-individuated modern consciousness into a larger whole.[1]

My decision to treat VR, in part, as a machine to realize such desires for bodily transcendence[2] is not intended to promote any particular metaphysics, though I do believe that many current materialist analyses of the technology miss the mark in failing to address the implicit importance of metaphysics to virtual consumers. Although military advantage, followed closely by global financial and data services, drives VR's invention, appeals to metaphysics, however subtle, remain important in promoting the technology. Such appeals would fail if they did not tap a pervasive cultural longing. Key VR inventors themselves evince various aspects of this yearning — often cloaked in a belief in progress. Eric Sheppard (1993, 4, 12) argues that information technologies are composed not only of machinery but also of the institutional and intellectual infrastructures that invent, deliver, and package them. What follows tries to keep Sheppard's caveat in mind. I offer a necessarily selective and critical review. Broadly speaking, I am interested in the ontology of representation, but I am also arguing that the form of a technology relates directly to perception — always culturally inflected, but only partially so — and to how ontology is discursively positioned. This precludes extensive discussion of *every* electronic technology (for example, TV), though I do address pre-twentieth-century technologies that influence the forms of both TV and VR. Finally, I agree with David Depew (1985) that history is criticism. A narrative history of VR is somewhat ironic given the technology's tendency to foreclose narrative/time in favor of spectacle/space, a consideration taken up in this chapter's discussion of science fiction.

Early Flight Simulation and Computation Devices: The Beginnings of VR Technology

Almost as soon as World War II began, the U.S. government initiated funding of flight simulator development. Research was arduous, yielding truly successful results only in 1960, just in time for the American space program. Yet by 1940 it had already been more than a generation since the first major air accident had occurred, in 1908, during a trial flight for the American War Department. Flight's power and danger made a flight simulator training machine desirable, and designs had been patented as early as 1910 (Woolley 1992, 42). In 1930 Edwin Link patented the Link trainer.[3] The pilot entered a mock-up cockpit equipped with controls through which a plane's pitch, roll, and yaw could be mimicked. Link's machine, with its pneumatic devices and hydraulic servomechanisms, was sufficiently evolved to imitate movements experienced in flight, as well as the sensation or force transmitted through physical contact with the joystick.

As Benjamin Woolley recounts, during the 1930s, research conducted by MIT professor Vannevar Bush led to a breakthrough in mechanizing the differential equations that were to allow the mathematical modeling of flight. During World War II, Link and others worked to physically reproduce Bush's mathematical model, and to marry the promise of Bush's differential analyzer (an early analog computer) to the basic physics of simulation. As applied to flight simulation, the initial challenge they addressed was integrating the "north-south" movement of the joystick with its "east-west" and up-down movements in such a way that moving the stick between any two compass directions would afford the trainee an adequate simulation of the resistance experienced in actual flight.

During this same period, designers improved the illusion of what a pilot might see from the "cockpit's" windscreen. However, adequate simulation awaited invention of digital computation and its ability to process the complex algorithms on which the "mechanics" of simulation rest. The Electronic Numerical Integrator and Computer (ENIAC), unveiled in 1946, was developed at the University of Pennsylvania as part of the war effort to automate production of the complicated ballistic tables required to predict missile and bomb trajectories. It was soon grasped that ENIAC might provide the advanced computation necessary for simulating flight, and that digital computers might exemplify a new technology that mathematician Alan Turing would soon identify as *universal machines*. Such meta-machines would render it "unnecessary to design various new machines to do various computing processes" (Turing 1950, 441).

Bush, who served as wartime director of Franklin Roosevelt's Office of Scientific Research and Development (Nelson 1972, 440), also theorized *personal* computation, but as a hypertextual extension of the self. His choice of words is strikingly similar to the description of the human-machine interface today called the cyborg. Bush's machine infects and enhances the human body and is predicated on the organic electrical dynamics of this body for its functionality. In his discussion "Memex instead of Index," Bush writes:

Consider a future device for individual use, which is a sort of mecha-nized private file and library. It needs a name, and, to coin one at random, "memex" will do. A memex is a device in which an individual stores all his books, records, and communications, and which is mechanized so that it may be consulted with exceeding speed and flexibility. *It is an enlarged intimate supplement to his memory.* . . . In the outside world, all forms of intelligence, whether of sound or sight, have been reduced to the form of varying currents in an electric current in order that they may be transmit-ted. Inside the human frame exactly the same sorts of processes occur. *Must we always transform to mechanical movements in order to proceed from one electrical phenomenon to another?* (1946, 32; emphasis added)

In 1944, researchers at MIT's Servomechanisms Lab, using digital equip-ment akin to ENIAC, successfully demonstrated that a light-sensitive, handheld detector wand, when pointed at a television-like screen adapted from radar technology, could select or "highlight" individual dots pre-programmed to move like bouncing balls across its surface. Such action bore similarities to reaching out to touch or contact an object. Through applied mathematics, MIT scientists simulated people interacting with concrete things, in the process unsettling distinctions between symbol and referent. The experiment suggested further investigation into human interaction with simulations. Other research (Bush 1946; Weiner 1948; Turing 1950) strengthened the idea that human-machine interactivity created a hybrid, "an ambiguous boundary between humans and inter-active 'intelligent' machines" (Biocca 1992a, 8).

Bush's "Memex" was a blueprint for a new technology culled in part from synthesizing existing devices and in part from his imagination. His description stimulated the scientific imagination in a manner similar to that achieved by the science fiction writing examined hereafter. Bush's imaginative contributions, and those of science fiction, are the fuel needed by spatial technologies such as virtual environments, which arguably de-pend as much on speculative narrative for their inspiration and genesis as on identified needs. The process is ongoing. Bush's writings resonate with recent contributions by computer scientist David Gelernter. Gelern-ter also stimulates cultural and scientific imaginations, but about possi-ble future virtual environments:

The picture you see on your display represents a real physical layout. In a City Mirror World, you see a city map of some kind. . . . You can see traffic density on the streets . . . the current agenda at city hall . . . crime condi-tions in the park . . . average bulk cauliflower prices and a huge list of others.... Pilot your mouse over to some interesting point and turn the altitude knob. Now you are inside a school, courthouse, hospital or City Hall.... Meet and chat (electronically) with the local inhabitants, or other Mirror World browsers. You'd like to be informed whenever the zoning board finalizes a budget? Leave a software agent behind. (1992, 16–17)

Both Bush and Gelernter borrow heavily from the technologies of the day — file cards and magnetic tape; computer screens and the mouse — and extend these in novel directions. In outlining their futurology scenarios, both men, pivotally involved in theorizing intelligent machines and cybernetics, point to a mind that longs to become a computer, one able to author a rational vision that encompasses all viewpoints. If only such a mind could take in enough information, it might finally realize the Renaissance wish to be "the measure of all things." There is a long history of equating the mind with the latest technology. Once the mechanical clock prevailed; today the computer, leading to metaphors of machine as man and man as machine. This is fodder for the science fiction that inspires those scientists writing virtual worlds into being, suggesting the instrumental power behind the "thrilling and horrifying possibility that we will someday bestow the sacredness of life upon matter, and the concomitant profanity of inert matter upon ourselves" (Pollack 1988, 21).

In Gelernter's City Mirror World, "lots of information is superimposed" on the display with which the user interacts. His comments dovetail with Terence McKenna's prediction that the "ambiguity of invisible meanings that attends audio speech [will be] replaced by the unambiguous topology of meanings beheld, [that] we will truly *see* what we mean" (1991, 232). Reduction, McKenna seems to say, is revelation. To layer information over sight assumes that information will become fully *known,* that through a technical apparatus operating like a philosophers' stone for the "information age," information can be transubstantiated into directly perceived knowledge that somehow bypasses the very mediation that is part of its production.

Such quasi-magical thinking also speaks to the contemporary Western subject's fear of being lost in space, or at least lost in the world. These comments imply a deeper yearning: that somehow, if enough information could be layered over our already media-saturated experience, if the mind could become a computer, could somehow recursively become the material symbol of itself, only then would the lived world become clear, "unambiguous," and fully understood. With the death of faith, only when we fully bestow on matter/machine the values previously attached to mind will we have the full resources to deal with what comes at us. However, in a culture increasingly reliant on visually dependent simulations of reality, what is beheld has also become highly ambiguous for the abstracted Western mind. Not only is it difficult to trust (*extend* belief) that, say, an unknown photograph has not been digitally remastered into something very different from what the camera first captured, but also the epistemology inherent in McKenna's and Gelernter's argument seems resistant to trusting actual physical experience. They would substitute the time and motility required to gain knowledge with fragmented, telecommunicated, visual representations of space.

Inventions and Cultural Forces

For VR theorist Myron Krueger, the computer's rapid evolution compared to that of earlier technologies — contrasted with the *lack* of evolution of the human form — leads him to theorize "that the ultimate interface between the computer and people would be to the human body and human sense" (1991, 19). The exponential enhancement in computing capacity from the 1950s onward — a key factor in making VEs conceivable today — is part of a "package" of long-term cultural and technological changes. The manufacture of the stereoscopic display, for example (discussed more fully in chapter 2), is a necessary development. The stereoscope and its modern entertainment and informational descendants such as the Viewmaster and stereoscopic photography are based on separate dual images, each depicting "the same scene from slightly different perspectives corresponding to human interocular distance" (Rheingold 1991, 65). When these are presented separately to each eye, our visual sense merges the two views into a single 3-D scene.

Edwin Land's work with light-polarized lenses is a separate precursive development that advanced the apparent cohesion of stereo images and was necessary for the creation of color 3-D film. Mid-1950s Hollywood features such as Alfred Hitchcock's *Dial M for Murder* required viewers to don special glasses to perceive the hallucinatory effects. In the film, the scissors that Ray Milland uses to menace Grace Kelly seem to fly forward from the screen to threaten the audience too, disrupting the "traditional" spatio-emotional remove that informs the relationship between viewer and screen. Although these 3-D experiments were cumbersome and were abandoned following Twentieth Century–Fox's more successful 1953 launch of the short-lived wide-screen anamorphic system dubbed Cinemascope, the ways in which they manipulate the spatial relationship between image representation and human perception to more directly involve audiences with the images before them is a conceptual progenitor of the computer-driven "ultimate interface" that Krueger describes.

During the same period, in an effort to woo back customers lost to TV, Hollywood experimented in heightening realism by stimulating moviegoers' nervous systems with atmospheric smells. Aroma-Rama, developed by Charles Weiss, sprayed "Oriental scents" through the air-conditioning systems of auditoriums where the documentary film *Behind the Great Wall* (1959) played. Smell-o-Vision piped odors directly to individual seats; it was used only once, in Mike Todd Jr.'s 1960 *Scent of Mystery* (Katz 1994, 52; 1263). Entrepreneur-inventor Morton Heilig's 1956 Sensorama Simulator offered the sensation of real experience through the multi-mediated use of 3-D images, binaural sound, and scent. It was influenced by the Cinerama process (an even more wraparound competitor of Cinemascope) and may be imagined as an individualistic precursor of the IMAX, IMAX/OMNIMAX, and IMAX 3D installations at science parks, museums, and, more lately, themed locations such as the Luxor Hotel in Las Vegas. Cinerama's extended horizontal projection reintroduced techniques developed for panorama painting, abandoned around 1900 following the cinema's enthusiastic mass reception. Vitarama — the predecessor of Cinerama — had been developed in 1938 by Fred Waller, who assembled five cameras and five projectors as part of a pilot training device that projected larger images on a curved screen (Oettermann 1997, 88). The inspiration Heilig derives from Cinerama, therefore, also extends to the military flight simulators noted earlier in this chapter.

Seeming to foresee immersive VEs, Heilig suggested in 1953:

> The screen will not fill only 5% of your visual field as the local movie screen does... or the 25% of Cinerama — but 100%. The screen will curve past the spectator's ears on both sides and beyond his sphere of vision above and below. In all the praise about the marvels of "peripheral vision," no one has paused to state that the human eye has a vertical span of 150 degrees as well as a horizontal one of 180 degrees.... Glasses... will not be necessary. Electronic and optical means will be devised to create illusory depth without them. (1992a, 283)

Heilig anticipates Mark Dery's observation that "in virtual reality, the television swallows the viewer, headfirst" (1993a, 6). Heilig continued work on his concept, in 1960 patenting his "Stereoscopic Television Apparatus for Individual Use," a "head-mounted display that a person could wear like a pair of exceptionally bulky sunglasses" (Rheingold 1991, 58).

Heilig's work in sensory immersion remained marginalized, in part because it was located within an entertainment milieu, in part because of lack of funds (Heilig 1992b). VR theorist Brenda Laurel (1993) writes of the conceptual breakthrough at MIT's Media Lab, where in the late 1970s and early 1980s researchers became aware of the qualitative difference induced when an individual sensorium was *surrounded* rather than — as with film, TV, and video — facing a screen at a distance. She finds that the vanishing *interface* this implies "broke new ground in bringing our attention to the nature of the effects that immersion could induce" (204).[4] Heilig, however, had already covered this "ground," unfortunately for him at the wrong place and time. His earlier and obscured entertainment-oriented research previews the Nintendo-directed escapism of current VR arcade games such as Dactyl Nightmare, in which pterodactyls swoop down through an illusion of 3-D space to snatch unwary players engaged in killing one another, carrying them "high" into the air, and dropping players to their virtual "deaths" on the cartographic chessboard surface "below."

Heilig's creativity notwithstanding, Ivan Sutherland is generally credited with synthesizing the trajectory now followed in simulations research (Krueger 1991a; Woolley 1992; Biocca 1992a). Affiliated variously with MIT, the cybernetics think tank at the University of Utah, and the federal Advanced Research Projects Agency (ARPA), Sutherland's 1965 meditation on virtual affectivity — "The Ultimate Display" — anticipates subsequent research and development of VEs.

> A display connected to a digital computer . . . is a looking glass into a mathematical wonderland. . . . There is no reason why the objects displayed . . . have to follow the ordinary rules of physical reality. . . . The ultimate display would . . . be a room within which the computer can control the existence of matter. . . . Handcuffs displayed in such a room would be confining, and a bullet displayed in such a room would be fatal. With appropriate programming such a display could literally be the Wonderland in which Alice walked. (1965, 506–8)

It would seem that metaphors of control, violence, and transcendence underpin such a wonderland from the moment of its conception.[5]

New media are assemblages informed first by the technologies and conventions of the past. Sutherland writes that "the force required to move a joystick could be computer controlled, just as the actuation force on the controls of a Link Trainer are changed to give the feel of a real airplane" (1965, 507). Sutherland models his ultimate display on flight

simulation. This is not surprising, for he also founded Evans and Sutherland, a leading flight simulation company. Given his prestigious vita of
computer science background and Department of Defense support, neither should one be surprised that this publicity-shy individual is called
the "father" of VR. Although Sutherland's genius helps make cyberspace
conceivable, military support for his work must be recognized—particularly following the surprise of Sputnik in 1957, and the swift American
response—both in the formation of ARPA, with its mission to synthesize
technological superiority and computational abilities (Brand 1987, 162),
and in the space agency NASA. Heilig's self-funded work had inverted
the "commonsense" temporal hierarchy often thought to exist between
military-industrial inventions and later socially diverting entertainment
spin-offs.

Sutherland's 1968 paper "A Head-Mounted Three Dimensional Display" accompanied his construction of a see-through helmet at the MIT
Draper Lab in Cambridge, Massachusetts (Stone 1992a, 95). Small TV
screens and half-silvered mirrors visibilized this early cyberspace VE. In
1969, at the University of Utah, Sutherland built the first HMD. It used
a "mechanical head tracker that fastened to the ceiling, called a Sword
of Damocles" (Brooks 1991, 11), and allowed a person to look around a
graphic room simply by turning her or his head. Two small cathode-ray
tubes (CRTs) driven by vector graphics generators provided the appropriate stereo view for each eye (Krueger 1991a, 68–69). Financed by
ARPA, the Office of Naval Research, and Bell Labs, the display marked a
step in realizing Sutherland's vision of "The Ultimate Display": "Our objective . . . has been to surround the user with displayed three-dimensional
information . . . objects displayed appear to hang in the space all around
the user" (Sutherland 1968, 757).

In resorting to HMDs, Sutherland and Heilig sought to go beyond
technical limitations of conventional film and TV that necessitate a space
between the technology and the viewer. The two men built on earlier
stereoscopic research, seeking to foster an illusion of three dimensionality.
Without stereoscopy, each eye would see the same flat, paintinglike scene
instead of one replicating the more "curved" sense of vision made available to perception via the slightly different position from which each eye
receives information and views the surrounding world. About VEs, Frank
Biocca notes, "we are not inside the space of the video image, only the
camera is. We are spectators, not actors" (1992b, 32). Biocca's distinction
is the same as one made by Thomas Hobbes between the citizen-Author
and the person-Actor. It is "we" authors who watch the performance of

the actors on the display or stage. As authors we have written both the actors and the stage. This distinction between spectator and actor also ironically parallels one made by Denis Cosgrove (1984) between the "inside" position of an individual experientially "in place," and an "outsider" who visually consumes a landscape set off by its frame. Extrapolating from Biocca, only the camera as metaphor for the disincorporated eye is at home *in* the image. Technology is needed to mediate between the two aspects of self that are present to each other.

HMDs influence the perceptually defined relationship Biocca notes between spectators and image, scene, or landscape. Using a tracking device connected to a computer, binocular vision and motion cues can be generated and continually adjusted to provide the sense of parallax that is one of the sensed "truths" or biases of our vision. Still, because of the physical weight of its auxiliary technologies, Sutherland's original display had to be suspended from above (Biocca 1992b, 37). Neither could it yet provide a truly emotionally real sense of the *surround environment* that was its inventor's goal. Two six-inch TV screens covered each eye of the user and offered him or her a stereoscopic computer-generated picture. Tracking sensors monitored individual position and movement in a partial reverse application onto the user's body of flight simulation technology's replications. Sutherland intended the objects in the computer-generated space accessed via the TV screens to be not only visible but *tangible.* He reasoned that the application of geometrical laws to reproduce size and shape could be extended to the application of physical laws to reproduce qualities such as mass and texture (Sutherland 1968; Woolley 1992, 55). The programming that lay behind the sense of resistance experienced through manipulation of a flight simulator's joystick could be applied to simulate the sensation of pushing and weight experienced with touch.

A brief excursus permits making a link between Turing's universal machine, how it comes to be applied, and Sutherland's "ultimate display." The latter exemplifies an aspect of Turing's machine — a concept depending on deductive Aristotelian logic to solve mathematical problems (Sheppard 1993, 3), and originating in a philosophy of mathematics associated with the symbolic logic advanced by Alfred North Whitehead and Bertrand Russell (see Bolter 1984, chap. 5).[6] Turing's abstract, immaterial machine is "a machine that can be lots of different machines" (Woolley 1992, 67). When personal computers (PCs) were developed in the early 1970s, their eventual poly-utility was recognized by only a few thinkers such as Sutherland and Ted Nelson (1973), Bush's and Turing's prescient remarks notwithstanding. Many "machines" reside within the PC, itself evolved from earlier single-use computational machines it has subsequently absorbed and displaced. The machine that processes the words on this page might also be used to design buildings, provide access to geographically distant information, or maintain the files and financial accounts of a commercial establishment. Such a comment is now commonplace. PCs can exist on their own or as networked into a communications matrix. But the PC can also be understood as an aspect of an abstract process of mind that has found physical expression. Jay David Bolter defines a Turing machine in this way: "By making a machine think as a man, man recreates himself, defines himself as a machine" (1984, 13). Computers are intended to be virtual machines in this manner. My screen simulates the thoughts I set down, but in its logic predicated on abstract mathematics, there is no necessary reason, as Sutherland grasped early on, why the machine need simulate only the actual or the real.

Turing's universal machine—a hybrid that can be many different machines, or "none of the above"—is a step in conceptualizing the electromechanical simulation of our selves, and one that partakes of a belief that to take the measure of all things is to *be* all things. In part, however, digital representations also achieve cultural status equal to or greater than their referents because they imply the universally coded standards that also reliably undergird stable political organisms. In the very naming of these devices and concepts— *ultimate* display and *universal* machine— there is a combined metaphysical and modernist suggestion of having come to an end or an irreducible element. The stability of a particular form of social relations is accorded a timeless universality.

Sutherland has made a separate contribution to the development of virtual worlds. In 1962 he developed Sketchpad, an interactive program that allowed a user holding a light pen to make designs on a screen that could be stored, retrieved, and superimposed atop one another (Sutherland 1963). Sketchpad—one of many products developed at the Air Force–sponsored Lincoln Lab at MIT—owes a conceptual debt to earlier advances in computer screen/human interface technologies developed for SAGE (Semi-Automatic Ground Environment), the 1950s command center controlling U.S. air defenses deployed against the Soviet nuclear threat (Edwards 1989). Operators monitoring the whereabouts of airborne Air Force planes used light pens developed from the 1940s detector wand noted earlier, touching the light pen to dots representing moving planes displayed on a computer screen (Manovich 1995a). As one outcome of SAGE's military agenda, Sketchpad further demonstrated that computers could be used for more than number crunching. The light pen transmitted the circle its holder first traced with the arc of his or her upper-body motion, and the computer simulated this action as a circular line on-screen. In a sense, the light pen (the precursor of the computer mouse) guided the human hand into a conceptual integration with the computer technology. If the "ultimate display" was a prototype of VR hardware, the earlier Sketchpad inaugurated a conceptual pathway for inscribing what would later be seen on the stereoscopic TV sets within the "display." It did so by reading human motion. No substantive training was demanded of the user. Pioneer computer and virtuality theorist Theodor H. Nelson has theorized Sketchpad's significance:

> You could draw a picture on the screen with the lightpen — and then file the picture away in the... memory... magnify and shrink the picture to a spectacular degree.... Sketchpad... allowed room for human vagueness and judgement.... You could rearrange till you got what you wanted.... a new way of working and seeing was possible. The techniques of the computer screen are general and applicable to everything — but only if you can adapt your mind to thinking in terms of computer screens.[7]

Nelson finds that "a new way of working and seeing was possible." The claim of novelty can be softened by noting that Sketchpad (or the mouse-screen-hand interface of a Macintosh computer or PC Windows environment) is a sophisticated way to sketch and erase — what anyone skilled at drawing would do on paper. What does seem novel, however, is that these technologies allow and facilitate a disavowal or displacement of authorship and artisanship. Users often feel less anxiety drawing with computers, as though they are not committing in the same way as they do on paper. Any novelty attributed to VEs must rely on the processes and relationships they engender (media as language), which may also be thought of as effecting a shift away from a set of moral attitudes toward creativity partly sustained by print technology. As N. Katherine Hayles argues, the body and the book are formed on a "durable material substrate. Once encoding [on either] has taken place, it cannot easily be changed... electronic media... receive and transmit signals but do not permanently store messages, books carry their information in their bodies" (1993b, 73).

Although innovations ranging from the computer mouse to text-graphics applications such as Hypertext and the World Wide Web are generally believed to be no older than their 1980s and 1990s commercialization, it was during the early to mid-1960s that these advances were first made at Douglas Englebart's ARPA-funded Augmentation Research Center in California. With the mouse, 3-D *gestural* input becomes a command language for computers. In Hypertext, users perform "automatic link-jumps" (Nelson 1972, 442) from one document to another by selecting specific icons on the screen. Collapsing and expanding multiple on-screen cut-and-paste documents is possible, as is using text-enhancing graphic imagery. All these functions imply virtual activity and conceptual leaps through representations of space that imply a "collapse" of space. All are precursive conceptualizations needed for realizing VEs.

Direct manipulation interfaces developed by Englebart and others at the Xerox Corporation's Palo Alto Research Center (XEROX PARC) remained relatively unused until Steve Jobs, Apple Computer's whiz kid, toured this facility (Rheingold 1991). If one seeks proof that Turing's ultimate machine lies waiting to be discovered within the imagination, Jobs's popularization of computing through marrying algorithmic power to graphic interfaces offers a fair example. Since the first Apples were marketed in 1984, applications have proliferated. These software packages are machines within a machine. They trade on the powerful merger between computation and the graphics programs computation makes possible.

XEROX PARC exemplifies a partial shift from military to civilian and business research that resulted from the Mansfield amendment drafted during the Vietnam War, as well as strategic long-term corporate diversification on Xerox's part. The Mansfield amendment limited ARPA funding to weapons-related research, yet in leading to certain scientists leaving ARPA-funded labs, it stimulated invention of personal computing by those who disagreed with U.S. foreign policy (Rheingold 1991, 85).

Nonetheless, ongoing research and development within the academy by those willing to accept ARPA disbursements remains central to the current development of VR. For example, during the early 1970s, the University of North Carolina at Chapel Hill (UNC) emerged as a major center in VR research, specializing in medical and molecular modeling and architectural walk-through or computer-aided design (CAD). The first graphic manipulator was created there. When its user moved a mechanical manipulator, a graphic manipulator on-screen also moved. If this image of a manipulator "picked up" another object represented on-screen, the user felt its weight and resistance (Krueger 1991a, 19). In the late 1980s, the Human Interface Technology Lab (HITL) opened at the University of Washington–Seattle. Connections with Boeing Corporation underline the virtual world's ongoing links with flight simulation. The GreenSpace Project, developed by HITL and the Fujitsu Research Institute in Tokyo, and partially funded by US WEST, is a virtual conferencing prototype. It has been used to link sites and individuals at the two research institutes. A separate video bandwidth is dedicated to transmitting facial expression. Photographically true facial representations are then merged into such idealized backgrounds as Mount Rainier and Mount Fuji.[8]

Contemporary VE research occurs within a complex and intertwined hybrid of profit-driven private consortia such as Autodesk, Apple, Division Corporation, and Silicon Graphics; entrepreneurial activity circulates between quasi-military facilities such as NASA's Ames Human Factors Research Division at Mountain View, California, and schools such as MIT, Brown, Carnegie-Mellon, Stanford, University of Southern California, UC Berkeley, University of North California, and University of Washington–Seattle. The 1970s move away from military applications is relevant, but NASA, despite the clear military implications of its space program, remains the engine of much research initiated by scientists who withdrew from the military orbit during the 1960s. NASA's mission has always assumed a taking leave of the earth. Accessing cyberspace — no less predicated on conceptually leaving the "space" of this earth than "cosmic" spaceflight — seems poetically congruent with NASA's broader mission.

Interbureaucratic rivalry also plays a role. Since the late 1970s, the U.S. Air Force, from research facilities headquartered at Wright-Patterson Base near Dayton, Ohio, has spearheaded HMD design. Under the direction of Tom Furness (who later moved to HITL) during the 1970s, a series of heavily funded projects into human perception and optics at Wright-Patterson led to the development of visual displays far more sophisticated than any then in commercial use (Krueger 1991a, xiv). These formed the basis for guidance systems used in the semiautomated American bombing of Iraq.[9] Wright-Patterson's placing of a million-dollar price tag on one of these displays in response to a request from NASA's Ames facility for a share in the technology spurred the less financially endowed Ames to devise its own display from existing technologies such as flat-screen CRTs. But NASA has gone much further than duplicating an HMD on the cheap. The DataGlove, originally acquired from the private company VPL (Virtual Products Limited), which later reengineered the glove for use with video games (Krueger 1991a, xvi), along with full-body input devices developed by a consortium of Ames and its subcontractors, more fully integrates the human form into virtual space than the earlier Air Force HMD. The sophisticated precision of the virtual war games technology the Air Force continues to develop and refine — one that has borrowed heavily from NASA's lead — illustrates the synergistic effects generated by this tax-funded competition between state agencies.

Visual perception theorist James J. Gibson[10] asserts that how we navigate our 3-D world and handle things within it determines and shapes our vision of the world (1966, chap. 13). Scientists at Ames, intrigued by his cognitive theories of the "perceptual invariance" by which he posits we see objects in the world, partly according to the "texture gradient" of these objects (an issue to which I return in chapter 4) sought to apply his ideas to the creation of VEs. The data glove that allows users to manipulate virtual objects both builds and simulates Gibson's belief that we grab on to our world and make it part of our "direct" experience. The extension of the user's virtual hand into cyberspace maps the dimensions of the virtual world on to internal human perception-structuring processes. Such mapping forms the basis of *telepresence* — "experience of presence in an environment by means of a communications medium" (Steuer 1992, 76). Telepresence, with its power to allow operators to influence and move material objects at a distance, allows a physical robot or virtual hand[11] to act as the servo-body of the person wearing immersive wraparound sensing mechanisms. The link between the human body and the robot is informational — remote control is at hand. The entertaining and transcendent possibilities of telepresence notwithstanding, this emerging technology is thought to be central to the engineering of a space station constructed by semiautonomous robots (Steuer 1992). But perhaps something more important is also at work here — the genesis of a belief in the body itself as only informational. This kind of reduction echoes Heim's description of cyberspace as a working product based on Platonism, one where "the dream of perfect FORMS becomes the dream of inFORMation" (1993, 89). Heim notes that virtual technologies, however, do *not* offer the direct mental insight Plato suggested would be available for true seekers of the light. Instead, such technologies clothe empirical evidence (the underlying binary software "engines") "so that they *seem* to share the ideality of the stable knowledge of the Forms" (ibid.). I would note that *information* is a series of rules and routines useful insofar as it is capable of being acted *upon*. With body-as-information, contingency and surprise are explained away.

Writing about the links between text, bodies, and VEs, Hayles notes that people "have something to lose if they are regarded solely as informational patterns, namely the resistant materiality that...has marked the experience of living as embodied creatures" (1993b, 73). Although she notes the potential loss, she also observes that interactions between people and these machines are increasingly based on exactly this kind of reductive patterning in which bodies materially correspond to a set of programmable signals.

As Hayles explains, "functionality" describes the communications modes active in computer-human interfaces. The hand motions of a data glove are one kind of functionality. However, functionality describes not only the capabilities of a computer but also how a user's senses and movements are disciplined to mesh with the machine's responses. On the user's part, VR technology demands a stylized hand. Humans build computers but are molded in return (1993b, 73).

Although VR may afford simulated access to a virtual and digitized community of representations—arguably a kind of "global public sphere" achieved at the loss of embeddedness and context—given the individuated manner in which the technology is being developed and will be accessed, the conflation between the conception it affords the user and this user's own perceptivity needs to be acknowledged and theorized. This conflation of conception and perception is more fully addressed in chapter 3; however, it should be noted here that, extending Heilig's Sensorama in surrounding the user's vision, the frame of earlier visual technologies such as landscape painting or TV recedes from view in VR and with it a degree of awareness of our separation from the machine. This receding of the frame, however, is already present in the panorama. Joining the idea of a receding frame to a positioning of the machine and user in close spatial proximity increases the potential for the user's active perception to collapse into the active conceptions contained within the technology. Part of an emerging "informational imperialism," immersive technology suggests that the conceptions it proposes are at one with the user's perception, thereby suggesting that the subject's independence is a fiction.

Like a braided desert stream whose channels rejoin downslope, the majority of the developments and institutional players noted in this chapter are brought together at NASA. Unlike most universities, NASA thrives on a kind of backdoor publicity (Rheingold 1991; Stone 1992b), and many of the writers most involved in popularizing VEs were permitted their first glimpse of cyberspace after hours or via a friend at Ames.[12] Not only has beleaguered NASA subtly publicized the cutting edge of its research in a way beneficial to its interests, but it also encourages use of virtual technology by medical and educational professionals. NASA's relatively open sharing of intellectual property (compared to the Air Force's, for example) is asserted to be for everyone's benefit.[13] One example will suffice. Time lag remains a problem in VEs. Overcoming its effects, as with motion pictures, depends on presenting framed snippets of reality at such a speed that they blend into a seamless illusion of realistic motion. But within VEs, the coordinates of the depicted space have to be recomputed each time a frame is changed, every thirtieth of a second. The demands of reality create a bottleneck for the current technology's relatively slow speeds by which multiple sets of commands and information must be simultaneously transmitted back and forth between user and hardware. NASA's cooperation with freelancers and small companies from several countries has meant that experimental computer architectures may receive NASA's financial support, for perhaps it will be an employee at a small software design company who will make a significant conceptual breakthrough. International private-sector research on transputers, much of it conducted by the British firm INMOS,[14] and partly sponsored by NASA and the British government, may succeed in refining computer architectures so that they become capable of juggling the vast ocean of data bits that run in parallel "pipelines" between the computer "platform" and the user interface devices such as HMDs, data gloves, and position trackers. Vast data-processing and transmission capacities are required to synthesize real-time eye-hand coordination at sufficient speeds to overcome the disorienting and reflexivity-inducing perception of time lag, an experience within which a trace of the user's subjectivity and of the constructed nature of the virtual world may still be found.

The Space of Science Fiction

> There is no more science fiction, there is only product development; and to some extent the reverse is also true. (Moulthrop 1993, 78)

Advances in computation form a pool of techniques from which virtual technology researchers can draw, select, refine, and redeploy. In its ties with the Western quest for transcendence — whether this be an out-of-body or off-the-planet experience — speculative "entertainment" equally sustains the will to develop VEs and is eloquently revealed in the pages of science fiction (SF).

Writing about advances in Geographic Information Systems technology, or GIS, during the past decade, Jerome Dobson comes close to connecting the dynamics of cultural technologies such as SF with advances in optical technologies. Dobson is aware that SF promotes widespread acceptance of technology. His writing gives evidence of SF's ability to suggest new directions for scientific inquiry and assemblages:

> In *Treasure*... a fictional search for lost collections of the Library of Alexandria... One character... laments, "Most geographic information systems are aimed North."... Technologists will snicker at this misuse of the term GIS.... But what if swarms of satellites were integrally linked to a central GIS... for the purpose of data acquisition?... Between... fiction and our reality, the principal differences are the degree of integration among systems, the speed with which computations and interactions occur, and the certainty associated with model results. (1993, 433–34)

In a fascinating study of the interplay of SF, postmodern academic theory, and virtual technologies, Scott Bukatman (1993) has coined the term "terminal identity" to refer to the birth of a new subjectivity at the interface of the body and the TV and computer screen. Within technology's increasing pervasion of concepts of the self, Bukatman identifies a growing belief that (hyper)individualism can merge with virtual technologies while current notions of humanity might somehow be retained without cost. Bukatman asserts that narrative form now gives way to spatialized concerns that engage our fixation with the proximities between embodied humanity and the electronic machines that facilitate an interpenetration of subjectivity and global capital flows.

Narrative has been a requisite artifact in the construction of the modern nation-state (Hobsbawm 1990; Anderson 1991). With postnational telematics, narrative gets in the way of data, and cyberspace becomes both the new spatial metaphor and the actual location of global power — one for which any isolated tech junkie might consider giving up his or her body in exchange for the fix of a wired fiber-optic communion "therein." With respect to making a connection between "the death of narrative" and the rise of VEs and spatial displays, Bolter (1991, 46) writes that nothing in picture writing corresponds to a first-person narrative because picture writing has no voice. For subjects positioned as pictures within VEs, there is space and other optical technologies but no *necessary* indication of the passage of time.

By constructing "a space of accommodation to an intensely technological existence," SF addresses how virtual technologies inflect our being in the world (Bukatman 1993, 10). Replacing visionaries such as Bush, SF is the new prescient mind that suggests a plausible image of the virtual world now under contract to be built. As the holodeck on the starship *Enterprise* in *Star Trek: The Next Generation* discloses, SF fosters a belief that technology might now offer humanity a wraparound alternative space to our present embodied existence.[15] The popular, if short-lived, TV program *Max Headroom* went one step further. Max is entirely relocated to within the digital space of interactive networked television. Fully sentient, computer generated, and the product of a human consciousness having been "downloaded" into an electronic net, Max communicates to others as an animated character on a screen. Viewers know that the world he and his embodied compatriots share is only "twenty minutes into the future." SF has always been an ideological narrative or "discourse." Its visions and overt use of spatial metaphors in describing power relationships offer compelling glimpses into the popular "geographic imagination" and are part of the apparatus facilitating actual technology's social acceptance.

Both Bukatman and Hayles argue that contemporary SF has turned away from its earlier interests in utopian futures and antipathy toward technology as the "other." They note the genre's discarding of conventional linear narrative in advancing the history of any one protagonist from one place within the story to the next. Narrative technique is superseded by descriptions of the merger of people and their technologized worlds.[16] People are less important than the technological systems within which they operate. The new SF is successful because it is honest to its thematic. Contemporary SF novels "embody within their techniques the assumptions expressed explicitly in [their] themes" (Hayles 1993b, 84). Such an authorial move could only be possible when "the posthuman is experienced as an everyday lived reality as well as an intellectual proposition" (ibid.) — a synthesis infusing SF from *Blade Runner* to *The Matrix*.

Heidegger's essay on the development of "World Picture" (1977) traces a Western belief that the world is best understood as if it were a picture in a frame *(Gestell)*. A picture such as a landscape painting relies on the technique of a bounded representative space closed to what our lived world might yet disclose. However plausibly it reads, SF promotes the idea of "world picture." Although the frame is removed in VR, one of the main uses of the technology is imaging pictures of the world — realistic or otherwise. To model VEs on SF is to load concept on concept, text on text. Although imaginative and creative, it suggests a feedback loop, like systems analysis, closed to exterior influence. A merger of people with their technologies would be a merger of people into concept. The "posthuman" has dispensed with the nonformulaic body in favor of codes, languages, and cultural productions — a contemporary and less hopeful reformulation of the story of Exodus as a flight from the supposedly "oppressive limits" of the body, followed by redemptive deliverance into the promised "land" of dataspace. Although I avoid psychological explications of the virtual world, as it is already sufficiently individuated without contributing to this further, Jean Starobinski's analysis of Sigmund Freud's theory of art as the expression of desires that refuse to seek fulfillment in the material world is provocative. Such desire is diverted toward the "realm of fiction" (Starobinski 1989, 139). Theorized in this manner, the deflection of desire onto both SF and virtual worlds takes on a "parallel" form of transference. Bukatman's treatment of the relationship between SF and VR is one of comparing two spheres that each gain in power by referencing the other.

William Gibson's *Neuromancer* (1984) is a science fiction vision widely acknowledged as having offered researchers following in the footsteps of Sutherland et al. a blueprint of the virtual world within the "ultimate display." "Gibson's cyberspace is an image of a way of making the abstract and unseen comprehensible, a visualization of the notion of cognitive mapping" (Fitting 1991, 311). It is also worth noting that this novel takes a popular understanding of physics and of Newtonian absolute space as exerting agency in and of itself and synthesizes these into a new metaphor of "cyberspace." Cyberspace is first created within computers and the IT networks that link them. Before any object might be inserted into cyberspace or represented "therein," a relationship must be established between or among spatially discrete computer terminals, or individuals relating with a set of interactive, graphic, spatial representations.

It is hard to overstate *Neuromancer*'s influence on the VR research community. Scarcely a thing written about VE and virtual technology neglects to pay the novel homage, and several VR business ventures are named after places and concepts in the book. Allucquere Rosanne Stone, one of the more considered theorists writing about VEs, argues that Gibson's novel demarcates the boundary between an information technology epoch extending from the 1960s until the book's publication, and the virtual reality and cyberspace epoch that ensues. She believes that this one novel

> reached the technologically literate and socially disaffected who were searching for social forms that could transform the fragmented anomie that characterized life in Silicon Valley and all electronic industrial ghettos.... Gibson's powerful vision provided for them the imaginal public

sphere and refigured discursive community that established the grounding for a new kind of social interaction. (Stone 1992b, 95)

Earlier I noted the fluid movement of virtual research personnel. Stone argues that the widespread background anxiety this movement promoted fostered a need that *Neuromancer* filled.[17] In modeling a plausible future for this spatially fragmented community[18] — defined as much by E-mail, computer bulletin boards (BBS), and listservs as any face-to-face geography — the novel gave voice to a nascent virtual community identity based on the alienated, insecure social relations within which researchers themselves labored. The novel, in turn, also suggested broad new avenues of research (Stone 1992b, 99).

Hayles's identification of two literary innovations deployed in *Neuromancer* — the "pov," or point of view, and cyberspace — can be read against Stone's thesis about the novel's appeal to spatially isolated researchers. These innovations "allow subjectivity... to be articulated together with abstract data" (Hayles 1993b, 82). According to Hayes, the "pov" is the mechanism by which individual consciousness "moves *through* the screen... leaving behind the body as an unoccupied shell. In cyberspace point-of-view does not emanate from the character; rather, the pov literally *is* the character" (83). Cyberspace is the datascape in which the pov can take place as a completely spatialized identity. In similar fashion to McKenna's and Gelernter's projects discussed earlier, awareness becomes pure vision joined to data. Data are humanized, and subjectivity computerized, "allowing them to join in a symbiotic union" (84).

The alteration of spatial relationships between viewers and what they see reflects parallel changes in technology and how it is deployed. With live theater, a viewer most often remains at a distance from the action on the stage, a distance reinforced by the proscenium. She extends herself imaginatively and emotionally toward one of the characters performing on a set that constitutes a fake space inside a real space. Yet despite Coleridge's dictum of a willing suspension of disbelief, the theatergoer maintains a critical distance of subjectivity within this temporary spatial relationship, a distance that is also expressed in "being present to oneself." In a similar fashion to the theater, early cinema actors played to the audience. Post–World War I American techniques, in contrast, progressively called on viewers to identify with the characters and their points of view (Manovich 1995b). "Accordingly, the space no longer acts as a theatrical backdrop. Instead, through new compositional principles, staging, set design, deep focus cinematography, lighting and camera movement, the viewer is . . . 'present' inside a space which does not really exist. A fake space" (Manovich 1995b).[19] Cinema permits a greater extension of subjectivity, along with a subjective narcissism, by suggesting that a person's self-interest may more fully lie elsewhere — actually "in" the set of a film — than her body's spatial coordinates.[20] In VEs, any vestige of distance remaining in film collapses into the pov, which is a transcendental perspective *and* an emblematic space that could only ever "take a stand" in the imagination. The frisson of transcendence and virtual control notwithstanding, an entirely informational (hence commodifiable) representation of the self as pov is made available for corporate and individual use (see figure 6, in chapter 4).

Earlier in this chapter, I critiqued McKenna's promotion of reducing all sensory meanings to visual topologies. The California sage of psychedelics and virtuality offers a most succinct and ahistoric understanding of the move from narrative to virtual spectacle and its potential impact on the modern subject implicitly repositioned by him as a picture.

> A world of visible language is a world where the individual doesn't really exist in the same way that the print-created world sanctions what we call 'point of view'. . . . if you replace the idea that life is a narrative with the idea that life is a vision, then you displace the linear progression of events. (cited in Rushkoff 1994, 58)

Cyberspace is first theorized in *Neuromancer*. For Hayles, cyberspace is this novel's second innovation: the immaterial space within which McKenna's vision will be played out. The novel also debuts Gibson's now famous "consensual hallucination" (W. Gibson 1984, 51) as one aspect of highly mediated social relations taking place in an intensely corporatized world predicated on overwhelming inequality and punctuated by a series of altercations between humans, human-machines, and machines that occur in material and virtual reality. The most cursory scan of writings on VEs makes clear that this concept has been latched on to with an astonishing tenacity, and that within the American VR community, consensuality has come to be equated summarily with equality. It is intriguing that so widely excerpted a concept has been so wrenched from the context in which it was located, for though the term is employed by Gibson, it refers to a polymorphous freedom less for individuals than for *data:* "Cyberspace. A consensual hallucination experienced daily by billions of legitimate operators. . . . A graphic representation of data abstracted from the banks of every computer in the human system" (51). For Peter Fitting (1992, 302–3), consensual hallucination is the novel's most striking concept. It demonstrates—even celebrates—the impossibility of direct, unmediated experience, for a consensual hallucination is always mediated and never subject, for example, to the ethics of face-to-face contact. In *Neuromancer,* it is also an experience in which the tension between positive and negative uses of technology has dissolved, along with the meaning or value of distinguishing between human and nonhuman—a duality already threatened (or promised) in 1962 with respect to human-computer interactions following the unveiling of Sutherland's Sketchpad.

"As if" Gods

Suggesting that science can inform fiction, and vice versa, Woolley argues that Gibson extended Sutherland's "looking glass into a mathematical wonderland" to the entirety of information. "With cyberspace as I describe it you can literally wrap yourself in media and not have to see what's really going on around you" (W. Gibson, cited in Woolley 1992, 122).

Although *Neuromancer* is used by academics and other cultural theorists in pointing to the bionic makeover of people into cyborgs, there is a general (though not complete) failure to note a broader thematic at work in Gibson's book. While the premise of the text has been interpreted as a radically dystopian consumerist future where "perception and experience are similarly contaminated" and paralleled with "remarkable new technologies and commodities [that] exist alongside the shabby and outmoded products they have replaced" (Fitting 1991, 301–3), it is the mutation of two corporate artificial intelligences, or AIs, into cybernetic gods that centers the real action. In humanist SF, such a change is always associated with a monster (technology = evil other), based on a dynamic similar to Bruce Mazlish's (1967) understanding of Victor Frankenstein's monster as technology spurned. It comes to pass that the humans must fight the monster/child/idea they have created. Whether humans win or lose, the battle between self and other is the primary moral locus that precludes any possibility of interchangeability between the two. Similarly, earlier research on AI was predicated on the frightening prospect of *replacing* human faculties (see Dreyfus 1992), a concept now somewhat superseded by what Heim (1993) identifies as a cultural theorizing of PCs as *components* of our identity—for some, a calming notion entirely consonant with current conceptions of the cyborg.

In *Neuromancer,* the protagonists, dimly aware that they exist within a society where embodied human integrity and history are passé, battle against establishment forces (the Turing Police) to allow this mutation from AI to god to take place. The eventual fusion of the two AIs — Neuromancer and Wintermute — is a meta-joining of both sides of a cybernetic brain by an evolved fiber-optic corpus callosum. Technology may be seen to use its creators to attain the state of union only technology itself has been existentially capable of imagining, thereby to achieve a kind of returning or unity. But this time the god is not the imaginary and therefore absolute and naturalized cultural technology of older religious belief but a systematic technology that humans have loved and set free. As Virilio (1994) notes, all technologies converge toward a deus ex machina: "Technologies have negated the transcendental God in order to invent the machine-God. However, these two gods raise similar questions."[21]

Gibson's AIs mutate into "a vast mind engulfing the whole of the Matrix. A god for Cyberspace" (Grant 1990, 47). If there is merit to Larry McCaffery's assertion that postmodernism is a condition that "derives its unique status above all from technological change" (1991, 3), then, as Glenn Grant argues, "if technology is to be our method of transcendence, Gibson seems to be saying, we should not be surprised to discover that our technology might have a greater potentiality for transcendence than we do" (1990, 47).

A less charitable understanding of the human-machine relations influenced by Western physics and technology is offered by Lewis Mumford. "Machines — and machines alone — completely met the requirements of the new scientific method and *point of view:* they fulfilled the definition of reality far more perfectly than living organisms" (1934, 51; emphasis added).

Neuromancer suggests that transcendence is to be achieved by machinic and virtual means. This argument is given weight by the novel's assumption that the human body will be "obsolete, as soon as consciousness itself can be uploaded into the network" (Stone 1992b, 113). In the aftermath of the novel's impact, the author has seemed less than comfortable with his creation, as expressed in the following self-parody of his own terse style.

> assembled word *cyberspace* from small and readily available components of language. . . . Slick and hollow — awaiting received meaning.
> All I did: folded words as taught. Now other words accrete in the interstices. . . . These are dreams of commerce. Above them rise intricate barrios, zones of more private fantasy. (W. Gibson 1992, 27–28)

Gibson's apologia may be an accurate critique of the ambition wedded to mathematical creativity within the virtual research community;

however, by 1992, cyberspace was no longer a concept awaiting meaning. The company Autodesk, for example, was founded by members of the scattered community Stone identifies who set out to build some of the novel's imaginative concepts. The discourse of cyberspace has been taken up by others, academics included, and reflects in part a "widespread desire to come to grips with the cultural implications of new electronic technologies" (Biocca 1992a, 17).

Neuromancer is the first of a trilogy that includes *Count Zero* (1986) and *Mona Lisa Overdrive* (1988). Although *Neuromancer* has been lionized to the extent that there is a "received truth" that the later novels cannot compare to its tour de force, the third novel, incorporating Gibson's critical recognition of the cultural processes *Neuromancer* helped set in motion, offers a more mature version of his VE futureview. I find the possibility of entering a cyberspatial *aleph* the most entrancing concept that *Mona Lisa Overdrive* details. Seemingly an amalgamation of Jorge Luis Borges's aleph, and German mathematician Georg Cantor's definition of transfinite numbers and theory of infinity, the aleph is "an approximation of everything" (W. Gibson 1988, 128), a place that is not a place, yet a complete synthesis of experience that feels as though it is. Cantor's work on set theory led him to posit that the cardinal number of a set of real numbers is larger than the aleph-null; in other words, the possibility exists for exponentially expanding worlds of mathematical irrational numbers to nest within even larger such worlds. Cantor's approximation is not dissimilar to Turing's machines within machines, or PCs and software. These technologies can be understood as "setting the stage" for the creation of virtual spaces that can "accommodate" the infinity of irrational numbers, instantiating "a physical theory of the irrational" (Porush 1996, 114). But Cantor was able to deduce from this what he called the power of the continuum, one that "is not denumerable, not algebraic, hence transcendental" (Reese 1980, 79; see also Porush 1996, 113–16). In *Mona Lisa Overdrive,* past fiction — allusions to Borges's magic realism — coalesces with the mathematics informing cybernetic theory.

The following two passages from *Mona Lisa Overdrive* provide a glimpse of the vision that the virtual research community has found so arresting. They suggest the ability of virtual technologies to fill a vacuum in meaning left by the explanation, and hence denigration, of the Christian God. In the novel's suggestion that virtual technology might fill this vacuum, an opening is offered to inventors and programmers who might themselves share in the power of creation and achieve a heady antidote to their alienated sensibilities. The first passage traces a succinct future "history" of VEs and might be read as a research agenda, or a comforting myth to virtual researchers that their endeavors will surely succeed, the "death" of narrative notwithstanding.

> *There's no there, there.* They taught that to children, explaining cyberspace. She remembered a smiling tutor's lecture in the arcology's executive crèche, images shifting on a screen: pilots in enormous helmets and clumsy-looking gloves, the neuroelectronically primitive "virtual world" technology linking them more effectively with their planes, pairs of miniature video terminals pumping them a computer-generated flood of combat data, the vibrotactile feedback gloves providing a touch-world of studs and triggers. . . . As the technology evolved, the helmets shrank, the video terminals atrophied. . . . (1988, 40)

Terminals that atrophy have already learned from their human inventors. The cyborg dynamic imbues "smart" machines and humans equally.

The second passage is a didactic exchange between one of the "fractured selves" (Continuity, an AI in the employ of the Sense/Net Corporation) contained within the cybernetic/alephic god and a human "construct" (Angie, a human modified with retinal cameras) seeking her cyborg origin and basis for identity. It is a contradictory blend of cautionary tale for, and tantalization of, the research community Gibson has helped identify. Continuity speaks first:

> "The mythform is usually encountered in one of two modes. One mode assumes that the cyberspace matrix is inhabited, or perhaps visited, by entities whose characteristics correspond with the primary mythform of a 'hidden people.' The other involves assumptions of omniscience, omnipotence, and incomprehensibility on the part of the matrix itself."
>
> "That the matrix is God?"
>
> "In a manner of speaking, although it would be more accurate, in terms of the mythform, to say that the matrix *has* a God, since this being's omniscience and omnipotence are assumed to be limited to the matrix."
>
> "If it has limits, it isn't omnipotent."
>
> "Exactly. Notice that the mythform doesn't credit the being with immortality, as would ordinarily be the case in belief systems positing a supreme being, at least in terms of your particular culture. Cyberspace exists, insofar as it can be said to exist, by virtue of human agency."
>
> "Like you."
>
> "Yes." . . .

"If there were such a being," she said, "you'd be a part of it, wouldn't you?"

"Yes."

"Wouldn't you know?"

"Not necessarily."

"*Do* you know?"

"No."

"Do you rule out the possibility?"

"No." (107)

The links between magic realism, transcendental beings, social relations, human agency, and VEs in Gibson's works are also found in Marge Piercy's (1991) foray into SF. *He, She and It,* set in the future and the past, suggests direct conceptual links between Jewish cabalist belief in the magical power of numbers and words and the contemporary belief in the power of software to create a separate mythico-spatial reality worthy of human attention and occupation. In parallel story lines, cabalistic ritual and AI research are used to create two artificial persons. Both the golem Joseph, created by the rabbi of Prague to protect his ghetto from an early-seventeenth-century pogrom, and the bioengineered android Yod, created by scientists to protect Tikva, a Jewish high-tech enclave that survives by producing encryption software in a future world of environmental degradation and corporate predation, are based on faith in the transcendental power of number and language, whether expressed in cabalistic incantation or software as the technologized Word. A discussion between two scientists anchors connections the novel proposes between cabalistic mystical doctrines and the scientific creation of an android, *as if* humans were gods, *as if* matter could transubstantiate to the ideal.

"You have trifled with the Kabbalah all the years I've known you," Avram said to Malkah. "Why do you bother? You're a scientist, not a mystic."

"I find different kinds of truth valuable.... In turning all statements into number, isn't gematria doing what a computer does? In fascination with the power of the word and a belief that the word is primary over matter, you may be talking nonsense about physics, but you're talking the truth about people."

"A person is as subject to physical laws as a stone is."

"But a person reacts and decides what's good or bad. For us the word is primary and paramount. We can curse each other to death or cure with words. With words we court each other, with words we punish each other. We construct the world out of words. The mind can kill or heal because it is the body."

"Malkah, politicians almost killed the human race by confusing saying with doing. . . . They confused the power of words over people with the power of words over matter — which is non-existent."

"You're making dichotomies, but in Hebrew the word *davar* . . . means word *and* thing, no distinction. A word, an idea, a thing. We see and hear the world with our minds, with words, in categories, not in raw sensory data . . ."

"You're becoming a Platonist, Malkah." (267–68)

Altered States

Several strands synthesize a virtual world. One is composed of hardware, software, and "wetware" — computer technology, human technical inge-nuity, and human bodies or components thereof. A second falls under the "cultural software" umbrella of "arts and entertainment." A third strand touches on the notion of transcendence, raised at various points in this chapter. The virtual turn by individuals such as the late Timothy Leary, prominent in the promotion of psychedelic drugs and up to his death a promoter of VR, speaks not only to a shift in interest from illegal to legal commodities as forms of release from material reality (as well as the progressive commodification of experience) but also to a continuing popular fascination with how symbolic forms of transcendence and magic might influence meaning and identity. Such a focus has broadened the interest in VEs beyond their military and entertainment applications.

Gibson's not quite dystopian future is as addiction prone as the pres-ent. In the novels, a mind-numbing array of legal and not so legal sub-stances is consumed by all manner and class of people residing this side of the interface in equal measure to the amount of time large portions of the population spend jacked in to individualized VEs Gibson calls "stims." Electronics penetrate the brain and mix with body chemicals. The resultant synergy on which stims rely is a brave new tele-vision in which users experience a somatic merger with the emotions and memories of cybernetically reconfigured media celebrities courtesy of electrodes im-planted at the base of their skulls. In the (very near) future, Gibson sug-gests, celebrity status will be divine.

In arguing the connection between virtuality and psychedelia, Mc-Kenna notes that "technology has already proven that it is the drug most palatable to the Western mind" (1991, 233). In a society of addiction, he wonders if VR will be judged a harmless substitute for drugs, but on a level more germane to this review, he notes that the synesthesia facilitated by VEs echoes the hallucinogenic reality where vocal performances are experienced visually and tactilely. Further, like the quality of a drug, the altered consciousness implicated by virtual living will be no better than the *quality of the codes*—the underlying software or language on which virtual living will depend and through which it will be conveyed.

Rheingold criticizes mainstream journalism's equating of VEs to an electronic LSD, noting that such writing deflects public attention from more real applications such as modeling radiation therapy for cancer patients, or walk-through CAD architectures (1991, 354). Such mainstream interest, Rheingold suggests, stems from a more general problem in American society about how to handle ecstasy, as in *ex-stasis*. In this, he supports a claim that transcendent imagining—wishing to enter a dream state—is one of the key drivers of this technology. He raises the possibility that people will use cyberspace to get out of not only their bodies but also their minds, and he argues the good of this by comparing VEs favorably to the inappropriate contexts within which real-time addiction takes place. An equally specious argument holds that methadone addiction is preferable to heroin dependence. Rheingold's focus seems to preclude any consideration of links between "the death of narrative" and the rise of visual technologies and their reliance on representations of space and addiction. Both addiction and the continuous circulation of images on view within VEs are narcotic stand-ins for embodied engagement and its greater emphasis on temporal continuity.

Addiction to images may seem to substitute temporarily not only for an older narrative form and the theoretical instruction it offered a slowly gestating modern subjectivity but also for the "traditional" or continuous aspects of places now discarded. Ironically, this continuity, today glibly dismantled as a limit or restriction on both individual freedom and the emerging global world of information "flow," provided the stabilizing counterbalance that anchored the process of change within which the modern self developed. To paraphrase Anthony Giddens, addiction now substitutes for tradition. With the decline of historical narrative and the continuing destruction of older places and communities by the forces of late capitalism, addiction, at least, is predictable: what is consumed is the same each time, the parameters become known, and surprise is minimized.[22]

2. Precursive Cultural and Material Technologies Informing Contemporary Virtual Reality

In the face of theorists writing critically about technology — Martin Heidegger, Lewis Mumford, Jacques Ellul, Herbert Marcuse, Marshall McLuhan, Langdon Winner, and Andrew Feenberg, among others — a pervasive cultural assumption, particularly in the United States, holds that technology is only a value-neutral tool. This precludes consideration of the social relations already factored into the technology by the scientific procedures leading to its development. It is more culturally reassuring theoretically to subsume communication technology under the metaphors of "medium" or "conduit" than to acknowledge any possibility of a technology's agency, however partial, contextualized, or inadvertent that agency might be — whether it results from unanticipated effects, applications, or poorly thought through research and design decisions on the part of a specific technology's makers. The "tool" approach concludes that communication technologies only mediate social relations, acting as containers or conduits through which meanings, social relations, and agents "pass" without being influenced by the passage. This begs the question "why invent a technology in the first place?" if it is somehow to be argued or believed that the technology — an assemblage of complex entities — has no power to influence or alter the state of the lived world, and the social relations contributing to and existing before the technology's introduction.

This chapter, together with the companion chapter that follows, examines VR within a number of historical and social contexts as part of a larger consideration of why the cultural desire to build VR technology exists. The positioning of VR as a *new* technology, the *next* thing, expresses a transcendental yearning to deny both history and the necessary limits that attend and organize material realities and their accompanying forms.

The extreme spatialization of VR, and users' relative ability, depending on the application, to reformulate the virtual environment at will, seems to confirm the efficacy of denying history as a narrative. At the same time, the technology's reliance on software and codes appears to affirm the social constructionist argument that "all the world's a text," our bodies included. If one concurs that the world and reality are always already socially constructed, then it would seem only "natural" to redirect a long-standing Western desire for transcendence toward technologies such as VR where limits appear to be constrained only by the imagination and the cultural contexts within which they operate. Nevertheless, VR technology does not emerge from a vacuum, and a number of social and historical considerations inform not only its built form but also the *idea* of virtual reality.

Modern individuals, charged with producing meaning and organized as the source of their own "truth," may well approach VR with desires to don or "perform" new identities as a kind of transcendence partaking of technology's power. VR promoters Sherman and Judkins suggest that the technology "is neither uncritically functional nor tackily quasi-scientific . . . it is poetic, mysterious, elusive" (1993, 38). Therefore, how I understand and theoretically situate the virtual technologies under review, as well as those earlier precursive technologies discussed hereafter — whether, for example, as merely value-neutral tools at one's disposal, or as sometimes able to attain a quasi agency, with the often inadvertent effects on social relations this may entail — is critical to informing any theoretical positions arrived at vis-à-vis these technologies.

To clarify my position on technology's relationship to people and places, I first consider the relationship between technologies and the humans who desire them built. Technological determinism and social constructionism are often positioned as antithetical, yet I think it more useful to chart a middle course between the two to consider more productively the social outcomes of technology, those both intended and unintended.

A delicate balancing act is at work in VR. Just as vision and sight have histories, an issue I discuss in chapter 4, so too do technologies: VR, an optical communications technology, extends aspects of earlier optical technologies, which themselves were variously positioned by contemporary designers and theoreticians. With this in mind, I discuss four prefigurative optical technologies and the expectations placed on them: the "perfect vision" of the camera obscura, the fantasy of the magic lantern, and the different immersive qualities of the stereoscope and the panorama. Their discursive positioning and repositioning over time inflects current optical practices informing VR. I discuss these technologies within their social and historical contexts, beginning with the medieval crisis of confidence and faith experienced by Christendom following the debacle of the Crusades, and I examine how Renaissance and Enlightenment understandings positioned and repositioned these technologies as required. Because of its lengthier history, and its being made a metaphor for different, even oppositional, theories of subjectivity, the camera obscura suggests ways in which VR extends and disrupts metaphors of light and vision, and relationships among viewer, subject, user, and machine. I introduce the magic lantern, stereoscope, and panorama to provide a context for a larger discussion of how these devices are reflected within the confines of VR, and to help explain why the technology has gained cultural acceptance as an *idea*.

With respect to more recent electronic technologies such as TV and video, it is true that virtuality, as a culturally mediated phenomenon, articulates with the media network within which Edward R. Murrow's "you are there" and Walter Cronkite's "and that's the way it is" came to be synonymous with broadcast journalism. Virtuality also reflects the quest for "remote seeing" that motivated early TV research as well as the quest for VR telepresence. I think it a more useful project, however, to interrogate and contextualize connections between VR and earlier nonelectronic technologies such as the panorama, which help prepare the viewer not only for cinema, *or* TV, *or* VR, but in different ways for *all* three. Although using different techniques, both panoramas and VR create 360-degree space. Like the camera obscura, both immerse the viewer in the technology.

Finally, with respect to issues connecting agency, technology, and politics, the works of Ulrich Beck (1992), Bruno Latour (1993), and John Searle (1995) are useful in understanding why technology has so often been seen as value-free and separate from politics and social relations. James Carey and John Quirk's (1969–1970) work on the technological sublime is valuable in informing my theorization of why virtual reality is being positioned as a utopian technology "beyond politics and language."

Technological Determinism versus Social Constructionism: Straw Man Arguments

To accord a potential of agency to technology implies contingency and change as factors escaping ironclad human control. There is an all too easy link between ignoring contingency and change and the "almost total disregard for the social consequences of technical choice" (Winner

1993, 439). Langdon Winner criticizes social construction's too rigid reliance on its "strawman: technological determinism," and the resulting failure fully to examine the "often painful ironies of technical choice" (446). He finds that social constructionists resurrect an old positivist "value neutrality" in the relativist guise of "interpretive flexibility" (447) when they argue that ethical issues raised by technology are undecidable because multiple readings of the "text" (in this case, technology) are possible. Or, extending Bruno Latour, social constructionists are really adhering to modernism's semi-explicit guarantee that science and the technology it produces — and politics and the social relations it influences — are mutually exclusive spheres.

David Rothenberg (1993, 14) offers a somewhat useful schema for thinking about how new technologies get introduced, one he calls "the circle of intent and result." Like the metaphor of the rhizome, the circle implies no beginning or end, but for purposes of explanation, I will begin with "human intention," say, a specific need or desire to communicate. Consider the telephone as a technical solution. As a technique developed by science and reflecting human intelligence and desire, it is put into the service of social relations. Its potential seems realized, but its usage, both as a thing external to ourselves and as a mechanism for extending our reach, then suggests new intentions. For example, there have been shifts in telephone deployment from short business communications to "keeping in touch," to accessing the Home Shopping Channel, to connecting the phone via modem to data-transmission facilities, and so forth. New uses harken back to human intention; from the user's newly expanded vantage point informed by telephony, new technologies are imagined by scientists and the broader sphere of social relations. Rothenberg argues that new technology helps realize existing human intentions and then inscribes these onto environments. "With technology, we turn the scenery into whatever we wish it to be" (16), a statement at least cognizant of the joint participation between science and social relations in human-wrought environmental change.

Rothenberg's circle of human agency and technical affect supports understanding of technology as an *activity* (Lyon 1994), as something that is done. Concerned about the extent of electronic surveillance, David Lyon notes that technologies often have capacities that solicit our use of them for surveillance. Critics whose field includes a technological dimension, yet who explain social relations without considering a relevant technology's role, are as myopic as technological determinists who claim that technology exists in a decontextualized social vacuum (44).

Technology is often understood as a "frozen moment" of social practice, not unlike the discrete events identified by the extremes of positivist and empiricist approaches. Machines, however, can also be thought of as arrangements of social practices in which the stability of definitions and relational positions between humans and machines of necessity must remain somewhat unstable and fluid. Asserting that not all actions are human, Donna Haraway acknowledges the difficulty of arguing this position given the naturalization of the dichotomy between nature and culture: "To speak of a world as a congerie of practices doesn't mean that all the actors are human."[1]

Michael Heim (1993, 77), writing about VR, distinguishes between interactive communications and tools. "A human user connects with the system, and the computer becomes interactive. Tools, by contrast, establish no such connection." Tools do not adjust to our purposes, except in a primitive sense. Yet the Cartesian metaphysics divorcing mind from body undergirds a view of technology as only an insensate object or tool that humans manipulate, rather than a process "that disrupts and reconfigures whatever we take to be 'essentially' human" (Markley 1996b, 6). In *Capital*, Karl Marx (1976, 492–93) rejected continuities between tool and machine, arguing that the latter is a complex entity and therefore always incorporates the history of its making. Machines replace specific skills of workers, and once set in operation, they use tools to do the same things workers did with tools. So, extending Marx, the computer mouse, for example, is a tool to the user and the machine, but VR is a technology — a complex assemblage of many devices and therefore not only a tool. Technology organizes laboring practices, in contrast to a person using a tool to extend his or her grasp or power. Using technology means accepting a hierarchy. If you don't know how the black box operates, you consent to a certain subordination (see MacKenzie 1996). Technology, therefore, is not only gadgets, mechanisms, and tools but, increasingly, also sets of social practices depending on distributed knowledge and skills (Miles and Robins 1992). These social practices, which interact with "the non-living physical world," are fundamentally and critically integrated with other social practices. "Instead of talking about the 'impacts of technology' we would talk about the co-evolution of technological and other social practices" (21). Miles and Robins further argue that new ITs based on digital telematics are distinct in their programmability (their ability to handle data in many different ways), and in their reflexivity (their ability to perform tasks and store and transmit information on how they have performed). I find these approaches useful in

thinking about VR and the virtual "world." Any examination of communications that focuses on social relations or technologies to the exclusion of the other skews results in the direction of social constructionism
or technological determinism.

Although the various "determinisms" — technological, social, and even
biological — are situated by their defenders and detractors as antitheses
of one another, all reflect the West's naturalization of causality, and all
"remain ensconced in the Newtonian world picture for which effects are
the direct result of an external mechanical cause" (Menser and Aronowitz
1996, 22). The linear progression of cause and effect, along with the notion of a unitary origin it instantiates, parses interdepending and interrelational processes into discrete events more amenable to scientific analysis, management, and control. A middle ground reserves a measure of
meaning to social relations *and* technology.

I noted earlier the widespread assumption that technology is but a
tool, and elsewhere that communications technologies are often positioned as simply conduits transferring messages between agents in a
neutral fashion, leaving the message intact. VR is both a technology and
a functioning communications environment or "medium." If, however,
it is seen only as a medium, then VR is more amenable to being accepted,
within a widespread cultural understanding that media are mere conduits between interacting people. What is implicit in the conduit analysis is that media do not really "mediate" but rather are only intermediaries,
without true ontological status, yet always distinct categories "between"
which a middle ground seems in short supply. For many potential users,
VR as a technology might denote an unpredictable, competitive "other"
with an agency of its own. Arguments that communications technologies are *tools* partake of a logic under which they extend our range, but
only human agency within specific contexts determines their value. Considering VR as a value-bearing technology means bringing to the fore
the formal constraints and hence biases to which it might be subject and
might help reify. Operating within these constraints, VR, for better or
worse, may also be thought of as a mediator, in the sense of its ability to
act, or to demarcate a space or middle ground, at or within which opposing categories might conjoin.

The position of technology-as-medium *does* acknowledge that humans invent technologies because they intend them to achieve certain
outcomes. Yet humanists and social constructionists, at loggerheads on
a variety of social issues, are often as one in dismissing considerations
of technology that would assign to it any agency or affect, treating such

considerations either as (1) examples of a myopic technological determinism blind to the social relations responsible for the technology's existence, or as (2) forgetting about what they often claim to be the most important point about theorizing technology — that it gets used in many ways and often for different purposes than intended. But this is obvious, as is the fact that a technology cannot be used for any purposes — its formal mechanisms both constrain and enable uses. Dismissing consideration of technological agency implicitly suggests that technological form per se ought not to be subject to critique. Under the sign of an overarching technical rationality, both criticisms and mystical accounts of technology alike are received with a sometimes self-serving skepticism that recognizes discussion of technology as jeopardizing a widespread (and naturalized) relationship between efficient machines and users — a relationship implicitly built on the narrative of progress within which science (and by extension its outcomes as technologies) is positioned as impersonal and cumulative (MacKenzie 1996, 217). This tends to leave the relationship between science practices and (for example) virtual technologies unexamined and in part flows from a conceptual evacuation of agency from all parts of the modern world other than the human.

One important consideration the discipline of geography can bring to this discussion is the role of scale. Both tools and technologies extend our grasp. Tool and conduit assumptions, however, do not address how an increase in technology's scale might increase its social effects. The root of the word "technology" lies in the Greek *techné*, which is a mode of knowing, a "practical rationality governed by a conscious goal" (Foucault 1988, 255), and "the name not only for the activities and skills of the craftsman, but also for the arts of the mind and the fine arts. *Techné* belongs to bringing-forth, to *poiësis*" (Heidegger 1977, 13). In *techné*'s connection to *poiësis* — a "bringing forth" manifested as the desire to make or build — it is caring about what one sees that brings forth this desire to build (Sennett 1991, xiii). Caring is foremost about what is at hand and in place, hence its applicability to the small-scale production of tools held in, and worked by, the hand. Contemporary insistence that premodern cultural meanings inhering in this linkage between *techné* and tools might still apply to how technology is theorized ignores established modern distinctions between scientific knowledge and its practical application as technology (Williams 1983, 315). A matter of degree can amount to a difference in kind. To blur the scale of a tool with the global scale at which technology now operates within networks diverts attention away from the manner in which specific kinds of agency and meaning have been systematically handed over to technology over several hundred years. Frederick Ferré (1995) offers this provocative example. It is possible, he argues, to adopt the view that the human mind is a computer. This is reflected in such phrases as "he was programmed to do it," or "why don't you process this a bit more?" However, it is also possible to define the self in contrast to computation. If a calculator multiplies better than I do — and comes to have a measure of what is considered to be intelligence — then I might not wish to continue to define the uniqueness of what it is to be human on the basis of "number-crunching and other symbolic manipulations" (128). Instead, I might choose to emphasize the emotional aspects of being human, precisely those aspects that computation eschews.

Whatever their scale, technologies operate in place. They reconstitute the meanings of places by becoming part of them, by linking them, or even by disarticulating them. The greater the technology in any one place, the greater its potential for affect, or transcendent power. It is hard to argue, for example, that should the tides of the Bay of Fundy be "harnessed" for their electrical power, we would then be able to suggest that the tides were a product of social relations, any more than we might imagine electricity issuing from the ebb and flow of the tidal bore without human mediation or intervention. Social construction forgets that humans are not the only agents in the world. Technological determinism ignores that humans creatively engage with the products of their own making in complex ways that defy simplistic reduction to cause and effect. As Menser and Aronowitz (1996, 21) argue, deployed technologies engage humans and the natural world such that the continuity or processual interrelationality and interpermeability that arise among all three preclude privileging or essentializing (the agency of) any one as a discrete event. VEs may have the potential to remap the subject's experience of self, but they will not do so without, at least, this subject's partial assent.

Smoke, Mirrors, and the Christian Eye: Casting New Light on "the Subject"

The roots of yearning for a virtual world are partly anchored by an ongoing Western belief in the eye as the most noble organ, and in vision as a sensual metaphor for *extending* understanding. This belief has helped set the stage for an emblematic virtual world of visibilized language that promises transcendence and affectivity in images, something denied us to date by our physical embodiment. Samuel Edgerton (1975) suggests that a shift toward the privileging of material vision (or sight; see chapter 4) as a metaphor for understanding and truth arose during the late medieval crisis of confidence and faith experienced by Christendom. The debacle of the Crusades helped set in motion a reevaluation of certain fundamental attitudes undergirding medieval Christian belief, including a demand for the development of a more powerful science to explain and conquer nature. As artifacts such as the Hereford and Ebstorf *Mappaemundi* (maps of the world) reveal, medievals possessed an adequate image of the world, though it was arguably a more "synesthetic" or place-inflected one than is the case today (see Barfield 1977; Ong 1977). The "purer," and ironically more explicitly cultural, form of vision based on the conceptual absolute space of Euclidean geometry that Roger Bacon (c. 1220–1292), for example, proposed was intended to provide a less sensually cluttered access to divine inspiration in face of loss by crusaders to the infidel. This loss was interpreted by Christian thinkers as resulting from a failure of devotional technique and the subsequent "faulty access" to God's instruction and command. Bacon's *Opus Majus,* written during the 1260s, petitions papal authority to redirect intelligent Christian inquiry and entreaty in accord with a *vision*ary perspective. Not distinguishing between vision and sight, Bacon places vision directly on an axis of truth and recommends elevating the status of geometry as a means of accessing

> the *ineffable* beauty of the divine wisdom ... [so that] after the restoration of the New Jerusalem we should enter a larger house decorated with a fuller glory. Surely the mere vision perceptible to our sense would be ... more beautiful since we should see in our presence the form of our truth, but most beautiful since aroused by the *visible instruments* we should rejoice in contemplating the spiritual and literal meaning of Scripture. (cited in Edgerton 1975, 18; emphasis added)

If divine wisdom — the Word — is unspeakable, then perhaps mortals instead might elevate the status of Logos's depiction. Bacon seeks to meld geometry with sight, representation with perception. His interest in "visible instruments" reflects the thirteenth century's great interest in optics and mathematics that followed the renewed influence of Neoplatonist thought, and its conception of space as infinite and open (Jammer 1969, 39).

In a sense, the history of vision in Western culture is a history of how sight has been colonized by mathematics, number, and various forms of Idealism. Geometry establishes a visual communication. Bacon wishes to enhance human perception with geometry to make it more divine, but in actuality he reduces sight to mathematics. As a practice, geometry blends "vision" and absolute space with an interest in deduction and logic as conduits to truth. For Bacon, depicting God's Word more purely through the use of representational geometric "picture language" is an abstract activity inheriting much from a Platonic vision of truth. Only when seekers of knowledge emerge from Plato's cave freed of their "corporeal shackles" can they attain the lucid and ideal realm of "active thought." In *Republic,* book 7, Plato (1987) would have it that all of us initially are imprisoned in a cave of ignorance, our backs turned away from the cave entrance as we watch flickering images or shadows on the back wall, which we mistake for truth itself. "The shadows are cast by objects being moved before a fire. The real world is outside the cave, containing the patterns from which the objects were copied, and the principle of the good, whose analogue is the light of the sun" (Reese 1980, 439). Only when seekers of truth and knowledge emerge from Plato's cave and into the "upper"(supranatural) world, freed from the "shackles" of this material, earthly reality for which the cave is a metaphor, can they begin to experience clear vision of real things present only to the "mind's eye" (Heim 1993, 88).

Bacon was not the first to recognize the power of merging vision and representation, as the much earlier exit from Plato's metaphoric cave makes clear. In his *Metaphysics,* Aristotle had temporized that "[Seeing], most of all the senses, makes us know and bring to light many *differences* between things" (Brenneman et al., 1982, 79; emphasis added). Privileging sight also privileges a spatialized understanding of difference: the break between our self and the world around us affords the best means available for accessing knowledge. Ptolemy's *Geographia*—a culmination of this early (and quite modern) geographer's efforts to represent a systematized relationship between the different features of the earth—is evidence of a second-century opticized understanding of the world, as are Al-Kindi's and Al-Hazen's theories of optics from the eighth and ninth centuries (see Lindberg 1976).

Heidegger (1977) notes that the propensity of visual perception is *curiosity,* a state of desiring inquisitiveness that may be contrasted to the more meditative state of *wonder.* Curiosity is of immense value to the analytical, logical Western science that Roger Bacon may be seen to call for, an inquisitive desire that later techniques such as Renaissance perspective painting appear to engage with and build on. Even today it helps direct the "shape" of virtual vision. Depending on emphasis in translation, *theoria* means either vision and/or truth, as in a "watching over" truth (Heidegger 1977, 163–64). Contemporary virtual technology is being refined to offer convincing "perceptual illusions." We live in a visual culture. As Biocca notes, when we seek information, "we '*look* into it.'. . . It is not surprising that a significant part of virtual reality development has tried to create better illusions for our eyes" (1992b, 30–31).

Bacon's papal correspondence is a call for what VR theorist and promoter Howard Rheingold (1991, 69) labels the age-old quest for "intellectual augmentation." However conceived, this enduring wish — which confirms to its holders that the here and now is never adequate — has called on a variety of communications practices in its quest to take on greater meaning and form. Today this means telematics or information technologies that synthesize telephony and digital computation. VEs, by blending visual communication with mechanisms that allow human gestures to be read by machines, form part of this will toward intellectual augmentation, which in the West has been defined as a *good* since at least the time of Bacon, if not Plato.

Bacon's papal entreaty offers this chapter a departure point for examining the diffusion of a progressively elevating status for vision in the West — an elevation that sets the stage for the visual communications technologies that follow.[2] At first these are print-based, alphabetic support for an individualized narrative of progressive selves. Yet in the current return to iconography for the depiction of information, there is an echo of a pre-Baconian and emblematic, medieval way of grasping reality. Both print-based and more purely visual "languages" or "picture writing," in which messages *seem* detached from words (see Bolter 1991, 46), depend on the science of optics and the eye. Yet picture writing's telematic manifestation may mark a partial return to a less linear, apparently more synesthetic grasping of experiential reality, suggesting an alteration in the relationship between human perception and communication. To achieve "intellectual augmentation," VEs, in concert with their underlying virtual technologies, propose we merge with the object of our infinitely curious gaze, which until now has kept us as modern subjects at its beck and call, alternatively enraging and tantalizing us to conquer it as an object, or worship it as a god.

Prefigurative Optical Technologies
Camera Obscura

It is customary to credit Renaissance Neapolitan Giovanni Battista della Porta with the invention of the camera obscura,[3] sometime before 1558. The thrust within American VR research is to have VR increasingly *corre-*

spond to the natural world—finally to achieve through technical means a "perfect copy" of reality that would be indistinguishable from that which it represents (see Bryson 1983; Coyne 1994). Belief in the eventual achievability of such correspondence, in part, extends and is informed by the dynamic underlying the Renaissance Doctrine of Signatures. This doctrine informs Porta's conceptualization of the lived world and his theorization of how the camera obscura might be used to represent and even double that world.

The Doctrine of Signatures asserted that imprints and signs are everywhere to be found in the natural world, and they reflect or communicate a use or intention that can be read and acted on. Such a belief is consonant with the contemporary understanding of medieval paintings, whose viewers understood them as animated by the real world of nature, of which everything, including paintings, formed a part. As I have argued elsewhere, for medievals, depiction in painting is literally true and understood as coeval with the material or imaginative reality being represented (Hillis 1994a, 4). Under the Doctrine of Signatures, referent and reference become the same; all things are linked regardless of time, place, scale, or (im)materiality. Yet although representation has since replaced the similitudes posited by the doctrine, it is a technology of thought to the degree that it renders *thought processes* more representational. Looking at the shape of a walnut's meat, for example, and linking this to the notion that walnuts must therefore benefit the brain—as many medievals believed—suggests a kind of dialectical thinking or juxtaposition on the part of the observer who gives meaning to patterns and shapes revealed to sight (see Manovich 1992). Walter Benjamin, in noting that the sphere of life that formerly seemed "governed by the law of similarity was comprehensive" (1979, 160), suggests that the ability or gift of producing similarities or what he terms "natural correspondences" is also the gift of recognizing them. Such correspondences awaken our mimetic faculty. They seem innate, and yet their manifestations reflect the cultural and historical contexts in and through which they take place. Extending Benjamin, looking at a walnut is not the same as staring into a computer-generated world if only because the imaginative engagement posited by the Doctrine of Signatures has been, in a sense, built or factored into VR's technological practices. Put another way, VR may issue from the awakening of mimetic imagination suggested by earlier forms of realism and correspondence because users may make links between what they see there and the material world. But Benjamin also suggests that the mimetic residue of similitude offers a resource against instrumental rationality and arbitrary language conventions. However, he also understands that "the rapidity of writing and reading heightens the fusion of the semiotic and the mimetic in the sphere of language" (1979, 162). This would also be a fusion of representation and similitude. In other words, though similitude may help resist instrumental rationality, its appropriation as a simulation within VR ironically confirms that a simulation of nature from which a user might imaginatively produce similitudes may actually reify the instrumental rationality that similitudes produced in other places might help resist.

Porta's description of the camera obscura anticipates the potential for VR touted by its promoters:

> In a dark Chamber... one may see as clearly and perspicuously, *as if they were before his eyes,* Huntings, Banquets, Armies of Enemies, Plays and all things that one desireth. Let there be over against that Chamber, where you desire to represent these things, some spacious Plain, where the sun can freely shine: upon that you shall set trees in Order, also Woods, Mountains, Rivers and Animals, that are really so, or made by Art, of Wood, or some other matter.... Let there be Horns, Cornets, Trumpets sounded: those that are in the Chamber shall see Trees, Animals, Hunters Faces, and all the rest so plainly, that they cannot tell whether they be true or delusions.... Hence it may appear to Philosophers, and those that study Opticks, *how vision is made.* (Porta 1658, 364–65, emphasis added)

In the modern world, the Doctrine of Signatures has been replaced by representation (Foucault 1970). If the Doctrine of Signatures was an earlier (and most imaginative) cultural technology that assumed a perfect correspondence of meaning between symbol and referent, or a similitude between things that looked alike, the progressive and exponential increase in the power of twentieth-century optical technologies to suggest the empirical truth of the illusion of reality they present not only supplants such imaginative conceptions but also, ironically, quickly works to confirm the *lack* of connection with earlier concepts and devices, thereby positioning recent technologies as truly novel. VR *is* a novel form of training ground on and in which users learn to overcome what would have been until recently resistance to the incoherent proposal that they might occupy the space of an image. This learning, however, is abetted by a lingering residue of belief in similitude: though today we claim to distinguish fully between images and referents, not only do users understand that images themselves are real but they may choose to allow the image to stand in for the reality it represents. In this, the technology's sophistication is critical; however, this choice made by users speaks to the essence of Baudrillard's *simulacra* and reflects the underacknowledged cultural capital still invested in *similitude*. Aspects of magical thinking are alive in such technologized practices of vision and optics. As I have suggested, however, so too is an ironic instrumental rationality at play. The aforementioned twentieth-century divergence toward data and away from physiology, however, could only have taken place within a mode of thinking privileging the eye as a *detached* optical device. And in this way, though VR implicates users' bodies (and thereby the material world), it also suggests a visual home for a disincorporated or excised optical subjectivity already present in Descartes's study of the camera obscura, *La Dioptrique*, published in 1637. Today, ironically, this subjectivity is prone to magical empiricism — one outcome deriving from the intersection of what has become for many a fetish for absolute certainty, combined with a belief that "seeing is believing," combined with living "within" and among a surfeit of overdetermined visual images, some of which refer to real things, others of which do not.

Porta's writing clearly shows his own synthesis of idea and technology, and his positioning of the camera obscura as both a scientific and magical device. Porta can be profitably read against VR theorist Michael Benedikt's description of the reflexive shaping of polyvalent "narratives" in VR. Benedikt hypothesizes a staggering increase in the future possibilities for rational communication in VR:

> You might reach for a cigarette that in my world is a pen, I might sit on a leather chair that in your world is a wooden bench. She appears to you as a wire whirlwind, to me as a ribbon of color. While I am looking at a three-dimensional cage of jittering data jacks, you can be seeing the same data in a floating average, perhaps a billowing field of "wheat." (Benedikt 1992b, 180)

The forms and cultural contexts of the camera obscura and VR differ, but each addresses an ongoing Western desire for transcendence from "this earthly plane," and each suggests that this might be obtained — if only virtually — through the illusionary fusion of images and reality, and abandonment of the embodied constraints of real places. Porta's camera obscura is a mechanism used by individuals for seeing and comprehending a shared external world given by God. Benedikt sees VR as allowing access to a subjectively given world that, despite his claims to the contrary, cannot be shared precisely because each user's world can be so different. In effect, Benedikt is proposing a reality that celebrates pluralism expressed as consubstantial *difference* along with a technology of representation that trades on the already-noted lingering or resurgent belief in similitude. Benedikt writes about a future technology, but good science fiction, as I discussed in chapter 1, always speaks to today. What Benedikt celebrates echoes Susan Bordo's (1993, 270) critique of the cultural practices of late-capitalist postmodern bodies: "A construction of life as plastic possibility and weightless choice, undetermined by history, social location or even individual biography."

If the camera obscura confirmed the Renaissance's belief in the consubstantiality of all things—a belief I am suggesting is ironically updated in Benedikt's assertions about the benefits of extreme polyvalency of form within VR—for seventeenth- and eighteenth-century Enlightenment thought, the same device was positioned as a model of visual truth, confirming the subjective interiority of viewers (Crary 1994). As a result, the camera obscura has been argued to inform photography's invention, and yet, although the two technologies bear many physical similarities, photography radically repositions the relationship between object and image away from authenticating the viewpoint and vision of the transcendental subject as a privileged form of knowing.

From looking at the relationship between camera obscura and photography, it is clear that assertions such as "VR is just a form of 3-D TV" are misleading. Such assertions facilitate social acceptance of *new* technologies without considering their implications, and they promote the wishful thinking that VR will be a utopian device. However, they leave unaddressed questions about its inherent reordering of subjectivity. Technologies are not neutral. As material components of ideologies, they help constitute as well as ritualize social processes and interests. With respect to VR, the technology mixes and matches Renaissance understandings of the camera obscura as confirming the equivalence of simulation and reality with an Enlightenment understanding of the device as confirming the truth of individual subjective vision—hence Benedikt's belief that the polyvalency of form in VR will augment communication by making each user's vision available to other users. Any communication in his model is between or among radically relative subjectivities who believe that total control over images-as-identities is key to a more direct communication with other images, machines, and (presumably) people. In other words, Benedikt hopes for a machine that delivers on the wish expressed in the phrase "If only you could see what I mean." Such a wish forgets that visual symbols and images, like language, are always culturally inflected and overdetermined. It also promotes the wishful thinking that VR as communicatory "space" would somehow obviate the need for discourse and negotiation of meaning.

Magic Lanterns, Panoramas, and Stereoscopes

The magic lantern, or phantasmagoria, as it was often called during the nineteenth century, was most likely invented in 1646 by a Dane (Godwin 1979, 83), though it is often credited to its popularizer, Jesuit Athanasius Kircher (1602–1680). This device used a projection booth in which an artificial light source was refracted by and through a series of lenses, each with an image superimposed on it. Light passes through the images and projects them on to a wall or screen (sometimes formed of vapor or smoke) in front of relatively immobile viewers, who, much as in the cinema, are in a darkened chamber between the projection device and the image.

The panorama was a 360-degree cylindrical painting, which, when viewed from the center, offered a sense of a simulated world that both surrounded the viewer and placed her or him at the center of its finite display. Designed by Irishman Robert Barker, patented in 1787, and receiving a successful commercial reception in London's Leicester Square in 1792, the device also provided a highly organized experience of spatial and temporal mobility. Unlike the experience of the magic lantern, viewers were required to move their bodies to see fully the "finite but unbounded" surroundings depicted on the circular perimeter of the space. Early versions featured painted landscapes of an earlier, bucolic countryside, which brought something of the country to the town, and of the past to the present (Friedberg 1993, 22). These landscape paintings themselves were modeled on actual landscaped vistas incorporated into country estates. Designed by individuals such as Humphrey Repton, these were bourgeois vistas of a rigorously reordered "natural world" (laborers included, on the other side of the ha-ha or invisible frame) and were designed to be viewed, if not entered (Cosgrove 1984; Daniels 1993). The receding of the frame, along with the resulting fetish for a realistic correspondence to reality — two of the features of most immersive VR applications and research — is already present in the panorama experience. The panorama "celebrates the bourgeoisie's ability to 'see things from a new angle,' [but] it is also a complete prison for the eye. The eye cannot see beyond the frame because there is no frame" (Oettermann 1997, 20).

Charles Wheatstone's invention of the stereoscopic display in 1833 revealed by instrumental means the importance of binocular vision in depth perception (Schwartz 1994, 40). The stereoscope and stereoscopic photography are outcomes of the sharp increase in the study of physiology taking place between 1820 and 1840. Using strategically positioned mirrors, the stereoscope is based on separate dual images, each depicting the same scene from slightly different vantage points, which together mimic the distance between our eyes and provide a sense of depth. David Brewster improved the stereoscope in 1849, enhancing its sense of 3-D photorealism. The Viewmaster is premised on Wheatstone's invention and, like VR, requires the viewer's physical contact with the device. When presented separately to a user's left and right eyes, the two disparate images are merged by the viewer's stereoscopic visual sense into a single 3-D scene. The stereoscopic display created within twin video display terminals (VDTs) built into contemporary head-mounted displays operates in a similar way in creating its illusion of immersion into virtual space.

Joining the Dots . . .

Although eighteenth-century discourse on the camera obscura positions a radical disjuncture between exterior world and subject, at base both it and VR are immersive. A space separates the viewer and the refracted image, but both are contained within the device. I would argue that this immersivity is what inspired Porta's pre-Enlightenment, premodern vision of "the light within." His vision is one of opening up the world to exploration and representation in novel ways. The notion of explorer/ exploration requires a freeing up of the encrusted medieval imagination that is, however, for Porta, not yet yoked to the Calvinist weight of responsibility that attends the individual's requirement to produce meaning. Jonathan Crary (1994, 38–40) argues that from the late 1500s, the camera obscura becomes *the* site of subjective individuation. The observer is isolated, enclosed, autonomous. From within the interiorizing and privatizing confines of the device, he or she witnesses the mechanical representation of an objective world and determines appropriate distinctions between this world and the visual representations inside the machine (41). This adjudication or politicized aesthetics flows, in part, from a desire to exclude disorder and privilege reason. The concept of a shared external world given by God is not so much rejected as it is supplanted by a growing awareness of an interior conscious (confirmed by using the device) increasingly focused on how it produces meaning and *orders* the world around it. The same technology that once confirmed God's plan now facilitates individuated perception of the world by the Cartesian *cogito,* and users may place themselves in a sovereign position analogous to God's eye. By the eighteenth century, the camera obscura will have been repositioned to confirm the superiority of an interiorized individual producing meaning on "his" own, fully in concordance with the Enlightenment's "discovery" that a light within the modern individual can be cultivated through reason, taste, and hard work (Taylor 1994, 27–30).

Crary (1994, 33) argues against making links between the camera obscura and the magic lantern. His important inquiry into the nineteenth-century subject opposes Enlightenment arguments about the camera obscura's relationship to an interiorized truth and a modernizing (Protestant) subjectivity against the Counter-Reformation context within which Kircher popularizes the magic lantern. Such oppositions, however, are always partially dependent on spatio-temporal contexts. Although Crary notes the centrality of all things optical to the twentieth century, his project does not specifically address how earlier technologies of vision collectively contribute, in various partial ways, to current optical inventions and processes. I have already suggested connections between the camera obscura and VR, and it is equally possible to theorize how the magic lantern spectacle prefigures the transcendent luminosity and the ghostly or "uncanny" illusions of today's virtual worlds. The nineteenth-century commercial success of magic lantern technology depended less on associations with divine illumination and more on a separate, sinister association with the spiritual. The deployment of magic lanterns for popular entertainment after 1802 — in contradistinction to how the technology was ostensibly positioned by scientists such as Sir David Brewster to dispel mysticism and the hidden mechanisms of illusion — *confirmed* the experience of specters, ghosts, and the spirit world (Crary 1994; Castle 1995). Interestingly, Brewster (1832) is disingenuous in this regard. He critiques the reliance on smoke-and-mirrors deceptions by pre-Enlightenment reactionary rulers and despots seeking to maintain power through fear and illusion (56–57), yet he is in awe of the contemporary smoke-and-mirrors technology. He not only describes how to construct a magic lantern but also exclaims over its favorable reception by a paying public (80–81).

If the camera obscura and magic lantern once reflected oppositional religious and ideological strategies of subjectivity and relationships of the subject to truth production, VR borrows aspects from any earlier optical technology that contains precursive mechanisms desirable to, and confirming of, a fracturing subjectivity seeking transcendence. VR thereby achieves a cultural point of purchase with such subjects who wish to maintain control over their individual production of meaning even as they play with the specter of abandoning the formal maintenance of modern identity to external sources such as VR and the "performativity" it encourages. Stated otherwise, VR is a world of images and data into which users insert themselves in search of greater productivity, enhanced subjectivity, transient escape, or combinations thereof. Whether positioned as a transcendence machine or as a utilitarian prosthesis enhancing thought, VR also reflects a desire for a return to either a prelinguistic or a prelapsarian state, or both (Hillis 1998).

VR, with its brilliant interior of images that may or may not (and need not) bear little relationship to the exterior world save for the socially inflected conceptions of software designers, clients, and users, is a world of artificial light; any "objective" world it models is contained within a computer program. The technology, therefore, not only sets aside the temporal hierarchy between outside object and inside image but also suggests that causal links between real-world references and virtual environments are less necessary than once might have been judged to be the case. Although VR dispenses with the dialectical model of clarity that Enlightenment thought believed was modeled by the camera obscura's relationship between exterior object (the real) and interior representation, by repositioning this object-subject binary entirely within its purview in a similar fashion to the panorama, VR appears to maintain the distinctions between the user and environment (or subject and space). However, it is also arguable that both panoramas and VR incorporate the subject into their purview, calling attention to the importance of scale in making such assessments. These distinctions, moreover, are culturally constructed and maintained. In the tradition of the camera obscura, the dialectic between self and world appears to be confirmed, even as image, language, and referentiality stand in for the real. So, VR maintains distinctions between an "anterior real" and its referents, even as it repositions these distinctions away from an observer who uses the technology to confirm these distinctions, to one in which users themselves are inserted into the dialectic of the technology to confirm the reality of the illusions it presents—images of themselves included. VR thereby suggests that the interiority, or "black box," of a computer program can operate in an adequate fashion to suggest an exteriority in opposition to users, who nonetheless, in Cartesian fashion, must imaginatively set their bodies aside to enter into a virtual world and, in a sublatory, almost re-medievalized fashion, merge with the display.

This ironic sense of merger also relies on the kind of immersive visual education provided in different ways by the stereoscope and the panorama. Crary (1994, 40) notes that the stereoscope advanced the conflation between real and optical. The reduction of the idea of vision this implies is wholly embraced by many members of the American VR community and is reflected in arguments that we will soon see what we mean. On a related note, like the stereoscope, VR relies on the saturation of gradients and surfaces users encounter with the kinds of visual details that Crary notes filled nineteenth-century stereoscopic images. A sense of flatness is at once confirmed yet denied through engaging the eye's attention with detail so that the implicit isotropic sense of space manifested in current immersive VR technology seems relaxed or given more of the attributes of a place experienced in extreme close-up. Unlike the panorama, movement in VR can produce a parallax effect. This quasi-hallucinatory or disorienting quality further deflects attention from the inherent flatness of the 2-D screen and images.

In her discussion of associations between early-nineteenth-century phantasmagoria technology and how usage of the word "phantasmagoria" has evolved, Terry Castle (1995, 141) notes that "something external and public" — the spectral illusions produced by the device — "has now come to refer to something wholly internal or subjective: the phantasmic imagery of the mind." The polyvalent world anticipated by Benedikt, wherein you are a ribbon of color and I am a jittering data jack, reproduces the modern belief outlined by Castle that we "see" figures and scenes in our minds, are haunted by our thoughts, which can "materialize before us, like phantoms, in moments of hallucination, waking dream, or reverie" (1995, 143). Belief that we see such materializations reflects the ongoing saliency of certain Stoic understandings of *phantasiai*—presentations or manifestations of what the soul seeks to see or believe (Goldhill 1996, 23). VR suggests the marriage not only of viewing and desire but also of its own externality (and the publicness that networked applications may provide) to the interiority of the human imagination "extended" to "engage" with privatized interior datascapes. Thus VR draws together an infusion of images, both sacred and profane, and of commerce and pleasure, all of which are generated within the machine.

In 1931, Benjamin wrote that "every day the need to possess the object in close-up in the form of a picture, *or rather a copy*, becomes more imperative" (1979, 250; emphasis added). As Benedikt's elegy to a kind of ironic hyperformlessness suggests, in a virtual environment, we each can be the controllers of our own phantasmagoria as we pursue individual combinations of "truths" once available via the camera obscura, escapes provided by the magic lantern, the panorama's sense of immersion in fantasy and new ways of seeing, and the uncanny sense of possessing familiar objects via manipulating their images in stereoscopes. Immersive VR further combines these kinds of pleasurable controls with the illusion that users might inhabit something like "the space of a dream" and coexist and comingle therein with copies of their "inner" thoughts, imaginations, and fancies. All of this relies on the play of light in virtual worlds. For the individual user, VR is interior illumination incarnate — subjective illumination conjoined to the machine in a hybrid or cyborg exorcism of interior subjectivity — even as the technology also confirms the light of a discrete inner subjectivity.

At the scale of VR's relationship to the politics of globalization and attending political economies, Stephan Oettermann's (1997, 45) assessment that the panorama was the art form of the Industrial Revolution is germane. European urban masses seeking thrills flocked to panoramas that operated as counterpoints to the dulling routines of nineteenth-century industrial and bureaucratic employments. Leisure operated in tandem with industry; by 1900, for example, sophisticated installations allowed viewers, surrounded by a false sky and vista of the sea, to stand atop a replica of a ship that pitched and rolled on hidden mechanical pulleys, wheels, and belts. With their 360-degree circular screens and direct and indirect lighting, fin de siècle panoramas anticipated 1950s Cinerama as well as themed semi-immersive experiences available in Las Vegas and other centers devoted to the secular religion of commodified leisure. With respect to leisure as a fully integrated component of commodity capitalism, the panorama experience has returned in updated form. The Ocean Dome, in Miyazaki, Japan, is a 300-meter-long, 100-meter-wide, 38-meter-high glass dome within which as many as ten thousand customers can frolic on simulated beaches, experience waves up to 2.5 meters high, and thrill to the effects of a simulated typhoon. The air is always thirty degrees Celsius, the water twenty-eight. Meanwhile, SSAWS (Snow Summer Autumn Winter Spring), near Tokyo, is the world's largest indoor ski run. Opened in 1993, SSAWS's longest trail is 500 meters, with a total drop of 80 meters. Fresh snow falls from nozzles in the ceiling and is kept crisp by a constant air temperature of minus four degrees Celsius. Like the panorama, both Ocean Dome and SSAWS feature elaborate themed interior environments. One difference is that the illusion is not depicted on interior walls, but the users themselves participate within a kind of hyperreal environment that takes up the entire interior space, whereas a railing or other device enclosing the central viewing space of the panorama established a more formal dialectical spatial relationship between spectacle and spectator.

If Ocean Dome and SSAWS owe a debt to the panorama, they can also be thought of as immersive virtual realities, predicated on control, and intended to provide an alternative to nature. Their experience is like living on the stage or having moved into the set of a film. They do not depend, however, on digital images of reality and the collapse of distance between viewer and a screen on which images are displayed. Rather, they are environments within which specific materials — water, sand, wind, snow — are technically manipulated. VR does not follow the rhythm of the Victorian factory machine of which Oettermann writes. The whirring cogs, even the vacuum tubes, have given way to software and pixelation. VR, however, does follow a rhythm — part of which harmonizes with the leisure economy, and with Ocean Dome within which consumers can experience a sense that technology will be the source of a sanitized natural environment it has otherwise worked to despoil. But VR's rhythm also echoes that of the maintenance of order through military power (the Gulf War and the Army Lab simulation) and of information technologies organizing workplaces and discursively repositioning the lived world as a global network.

Politics Divorced from Science

Despite the clearly embedded social contexts of technologies, belief persists in the myth of a value-free technology. In a brisk monograph, Latour (1993) argues that the genius of early modernism was to have created gaps between meaning, nature, and social relations. Yet he is explicitly concerned about how these gaps might now be narrowed. Latour uses the work of Steven Shapin and Simon Schaffer (1985) to trace the seventeenth-century origins of the disarticulation between science and politics, and the subsequent "canceling out" of the older Christian God. The political philosopher Thomas Hobbes and the natural philosopher and empirical scientist Robert Boyle agree about many things: "They want a king, a Parliament, a docile and unified Church, and they are fervent subscribers to mechanistic philosophy" (Latour 1993, 17). They disagree, however, about what can be expected from scientific experimentation. For Hobbes, the unity of the social contract under which the sovereign achieves the power to govern over all is threatened by the multiplicity of opinion represented by the methods Boyle uses to empirically verify the as of yet incomprehensible but nonetheless incontestable fact of the vacuum he has created by means of his mechanical pump. For Hobbes, this is untenable. It represents an independent opinion, empirically verified by the "witnessing" of a group of individuals and the unreliable sensory mechanisms on which they rely, instead of a universally agreed on and ideal calculation based on mathematics. Worse, from Hobbes's perspective, Boyle's experiment and subsequent empirical verification by others occur in the private space of the lab, over which the state has minimal control (Latour 1993, 20).

Boyle's construction of the vacuum pump, and having his peers witness its performance, reveals hard facts. These have been brought to light by human endeavor, yet they are not socially constructed. Boyle's vacuum exists apart from, and independent of, his own corporeal reality. John Searle makes a useful distinction in this regard between "institutional facts" — "facts in the world, that are only facts by human agreement" — and "brute facts" such as "snow" or Mount Everest, which are independent of human opinion (1995, 1). In the case of determining an institutional fact, process has priority over product (57). Language is essentially constitutive of such institutional facts or realities (9). I would note that though we perceive VEs sensually, they constitute institutional facts. They are socially produced but are being culturally positioned to masquerade as brute facts — a discussion further developed in chapter 6.

In time, facts such as Boyle's vacuum — though their existence requires human participation and verification — are accorded via an unacknowledged quasi-magical thinking the status of agents. This comes to pass, in part, because such facts are endowed with meaning produced through the use of language, but also because they are incapable of will and bias and therefore judged by an emerging empiricism as more reliable than human memory and recall. This contrasts with willful humans, who lack the ability consistently to "indicate phenomena in a reliable way" (Latour 1993, 23). However, institutional facts enjoy a renewed "expression of commitment" each time they are referenced or called upon by humans. "Individual dollar bills wear out. But the institution of paper currency is reinforced by its continual use" (Searle 1995, 51). The practice of independent scientific empirical observation, determination of, and peer agreement about such facts outraged Hobbes. He foresaw its eventual potential to undermine the delicate balancing act required to maintain peace under a social contract imposed by the state on its citizens, in which the rights of the people are limited to being represented by the state. A multiplicity of contestable opinions about the existence of facts confirmed by sensory verification and discourse would challenge the state's right to absolute representation over all. Uniqueness runs afoul of representation. Latour reminds readers of the original commonality between the scientific representation of facts (constant revision of models of reality, and known to us only through our senses), and a representational system of government; and of the genius contained in the subsequent splitting of the powers and responsibilities of scientific representation and political representation into separate, conceptually independent spheres. Hobbes's state needs science and technology, and scientific practice requires a strict delineation of the religious, political, and scientific spheres to flourish (Latour 1993, 28–29). The famous dispute between Boyle and Hobbes is not, however, only about divorcing science from politics and excluding religion from both.

> Boyle is creating a political discourse from which politics is to be excluded, while Hobbes is imagining a scientific politics from which experimental science has to be excluded. In other words, they are inventing our modern world, a world in which the representation of things through the intermediary of the laboratory is forever dissociated from the representation of citizens through the intermediary of the social contract. (27)

Within this modern political innovation, the "representation of nonhumans belongs to science, but science is not allowed to appeal to politics; the representation of citizens belongs to politics, but politics is not allowed to have any relation to the nonhumans produced and mobilized by science and technology" (28). The deeper contest between Hobbes and Boyle is about determining which resources will be available to the separately evolving spheres of the science of things (though constructed by "man"), and the politics of "men" (though sustained by things).

This separation of the spheres frees enormous energy for scientific inquiry and political innovation, but the spheres are culturally constructed and held apart for the benefits this confers. It is intellectual life that separates the realms of nature, social relations, and meaning. This separation is one of the "contradictions of modernity" (Sack 1992, 8), and also one of its great strengths.

For Latour, this separation—a purification of principles—has led to paradoxes and subsequent attempts to resolve them by what he identifies as a "constitutional guarantee" offered by modernism. Figure 3 reproduces the chart by which Latour (1993, 32, fig. 2.1) models these "paradoxes of Nature and Society." Reading it back and forth and diagonally suggests something of the underlying cultural assumptions by

First Paradox

Nature is not our construction; it is transcendent and surpasses us infinitely. (A)	Society is our free construction; it is immanent to our action. (B)

Second Paradox

Nature is our artificial construction in the laboratory; it is immanent. (C)	Society is not our construction; it is transcendent and surpasses us infinitely. (D)

Constitution

First guarantee: Even though we construct Nature, Nature is as if we do not control it.	*Second guarantee:* Even though we do not construct Society, Society is as if we did construct it.

Third guarantee: Nature and Society must remain absolutely distinct: the work of purification must remain absolutely distinct from the work of mediation.

Figure 3. Bruno Latour's "Paradoxes of Nature and Society." Adapted from Bruno Latour, *We Have Never Been Modern,* copyright 1993 Harvard University Press.

which technology, for example, remains conceptually off-limits for many social science and humanities workers, save for its being acknowledged (and deployed) as a tool. It also suggests why technologies are seen as mediations by humanists and social scientists. Such a reading can also be used to suggest many of the contradictions gathered with VEs, which act as an agent of meaning for scientific technology and as a theater of social relations.

Under the nature/culture dichotomy, statements A and B may be seen to agree, as may statements C and D. Beliefs or theories holding to the greater importance of immanence in the world would find that statements B and C do not contradict each other, whereas those theories subscribing to the greater place of transcendence in human affairs would find a similar result in holding together statements A and D. However, it is more likely that most would find statements A and C, as well as statements B and D, to conflict with one another. Yet all four statements enjoy varying degrees of cultural and, more specifically, academic and intellectual favor. Modern societies, segments of which hold to these four statements as something akin to tenets of faith, have managed to produce hybrid forms of nature/culture cultural technologies — such as VR — without too overwrought a concern for violation of the statements.

Although premodern cultures developed mythologies that crossed the boundary between nature and culture, most such cultures rigorously distinguished between the two. As a result, they tended to freight nature and culture so heavily with meaning that little room existed for developing the kinds of nonmythic hybridized understandings that lead to material procedures, technologies, and organisms. Such material hybrids, often unacknowledged, are a proliferating feature of modernity. In no particular order of ranking or scale, some examples other than VEs include development of hybrid strains of wheat and other grains, new reproductive technologies and genetic engineering, the concept of the cyborg shared by academics and science fiction alike, the understanding of what is a virus, the theory of light as both a particle and a wave, interdisciplinary studies within the academy, and the contemporary concept of "network." These modern (and postmodern) categories are more abstract and open to a variety of meanings than premodern ones.

Yet Latour (1993, 35) also notes a modern dismissal of the "obscurity" of the premodern period, which "illegitimately blended together social needs and natural reality, meanings and mechanisms, signs and things." This understanding of the premodern world reads like the theorization of place as the joining of meaning, nature, and social relations. Victor Walter (1988), in arguing for a reinvigorated theorization of place as an antidote to widespread environmental degradation and loss of meaning from the lived world, claims that he seeks from place a "radically old" way of knowing the world. Walter's call is neither antimodern nor postmodern, but one for a renewed appreciation of the best "parts" of what was lost in the modern purification of the world, which "cleanly separated material causality from human fantasy" (Latour 1993, 35). This "progress" came at the price of moderns being "unable to conceptualize themselves in continuity with the premoderns" (39). The continuity of places is sundered. What remains is nonstop actual change and the continuity of the conceptual purity of categories. It is intriguing that virtual technologies — with their ability to suggest an experiential collapse of scales of meaning, fusion of figure and ground, body/machine, subject/object, here/there, telepresence/copresence, and exteriorization of aspects of memory into programmable iconography — evince something akin to the premodern synthesis of places. It is as if to say that with the withering of this synthesis in the developed world's public places, an optics/language technology such as VR that "speaks" through light and mirrors, developed for military and commercial applications, has been found to have a potentially equally important function of suggesting something of this earlier synthesis now felt by many to be lost.

The potential of VEs to deliver such a synthesis notwithstanding, all cultures are interested in the purity or coherence of their operational categories. The genius of the modern West has been to suspend consideration of future consequences for the social order that flow from the production of nature-culture hybrids. Hybrids have been somewhat protected and insulated from the consequences, articulations, or effects they help produce. Because of this protection from danger — from too powerful a critique of science by religion, for example — the production of hybrids has proliferated. Yet because of the sanctity attached to the demarcation between science and nature, and politics and society — despite the increasingly numerous links between public policy and "science policy" — it has been difficult for the academy, itself increasingly specialized, to acknowledge fully, let alone study, the impure hybrid.

Difficult, but not impossible, as Latour's work demonstrates. So, too, does that of Ulrich Beck (1992). He argues compellingly that because science produces reality, and has done so for quite a while, it has therefore developed a set of historical markers and contexts that can be used to make the moral arguments that not only does science cause problems — in other words, is directly implicated in the political sphere — but that if it was once and still remains a taboo breaker, so too has science, operating under the tradition of modernism, become a taboo *maker* (158). Science as a taboo maker suggests a close proximity to the hybrids Latour identifies. Beck's work suggests that it could not be otherwise, and to avoid coming to terms with the consequences of what Latour terms hybrids only augments the proliferation of never before experienced risks unleashed by science and its industrial applications.

Within the technological "space" of a VE, representations of society are "freely constructed." This is the promise of "interactivity." Yet a VE also proposes itself as a representation of the natural world. A "virtual nature" composed of synthesizing concepts of absolute and relational space is constructed in a lab. Both society and nature are immanent within a VE at the level of invention and production. But at the level of individual engagement or consumption, society and nature both may seem transcendent. The interactivity allows users to behave as if they at least partially constructed society and nature, even though what is being constructed is a model that one interacts with according to its necessarily limited range of preprogrammed possibilities. However, the iconographic nature of the spatial display on which the virtual world takes place is predicated on offering a "direct perception" of reality, in contradistinction to the more abstract and self-reflexive modes of apprehension available to visual sensation via text. Users may feel as if they are directly experiencing reality, even though it is a model — a culturally specific subset of the reality it represents. VR is always of this earth, certain imaginative yearnings and conjurings notwithstanding. In Latour's terms, VR is a hybrid, but its claim is culture *over* nature and not a "culture-nature" or "nature-culture" fusion.

The military simulation described in the introduction is interactive. The computer program is able to represent to the user his or her continually evolving point of view, challenge this user with hostile agents, and permit the user to walk/fly through the environment. This contributes to a sense that the user does not (fully) control this world, notwithstanding the fact that it has been fabricated by humans. Yet in a parallel but inverse fashion, the institutional fact of the VE is also able to suggest that even though "we" do not construct society per se, users are (seemingly) able to construct society at will. Finally, however, distinctions between nature and society are somewhat set aside in this virtual theater of military preparedness. This is not so much because the simulations draw together town and country, or culture and nature, so much as the software and hardware permit the hybrid suggestion of an independent agency for a nature that is also representationally under users' control. VEs thereby somewhat maintain the nature-culture divide even as they fully mediate between the two spheres — *seeming* to produce something of the fusion suggested in Menser and Aronowitz's (1996) remedial and healing proposal of "natures-cultures." However, any seeming fusion approximating "nature-cultures" in VR is a risky *optical* illusion, as the "natures-cultures" virtual synthesis is entirely embedded within a socially constructed view of nature, one "made real" by the visual images a user sees in the stereoscopic display created by the video display terminals inside the head-mounted display. The integration of a user's body and the machine that takes place "in" the virtual world of a VE implies that human bodies themselves are parts of culture and therefore naturally can be conjoined with technology. An implicit claim is being made that there is no natural world, or at best that the "natural world," as is the case for explanations and representations of nature, is really a subcategory of culture. VR, then — despite its utilitarian applications — is not a value-free technology, a mere "conduit" designed to communicate "pure" information in a postsymbolic fashion; rather, it is always already inflected by social, political, theoretical, and ethical assumptions.

The Virtual Sublime: Desiring Utopia

The view that technology is nothing more than a tool partakes of utopian thinking, a belief that technology is part of an inevitable progress toward a future in which social ills can be managed, if not cured, by technological fixes. Such a view, of course, downplays the fact that technology can have various kinds of agency and influences outcomes, including both those anticipated and those unanticipated by the scientists, designers, and engineers who help move technology from idea to practice, from the door of the lab into the video arcade, hospital, or military training institute.

Writing in 1970, James Carey and John Quirk critiqued futurologists of the day such as Marshall McLuhan, Buckminster Fuller, and Konstantin Doxiadis for subscribing to a technological sublime that led to a belief that electricity would overcome the obstacles that defeated earlier utopias; a forgetting about a recentralization of power made available via the computer and energy grids; a failing to consider an erosion of regional cultures; and an ignoring of the emergence via media of a single "national accent" in tone and topical coverage. Reflecting the continuing salience of the technological sublime, VR reflects American cultural myths (though Japanese interest in the technology—termed "Intimate Presence"—must run a close second). In its blurring of technology and experience with a decidedly technical emphasis, VR connects to an America as Utopia for Europeans fleeing another, earlier kind of spatial tyranny. Carey and Quirk, extending the work of Leo Marx (1965), note that from the beginning, technology was welcome in the (American) "garden" (Carey and Quirk 1969–1970, 223; see also Nye 1994). This naturalization of technology supported a belief that machines could be used while the hierarchical excess of a (European) mechanized society could be avoided. Faced with the empirical evidence of American industrial slums and other products of the Industrial Revolution, weaknesses in the myth of machine in the garden were addressed by creating spatial zones outside of the industrial devastation. The first such zone was nature. When it fell, metaphysical spaces were created—pluralisms, tribal ethnicities, cosmic states, and so on. To this list I would add eventual quasi-metaphysical solutions such as VR, a technology and social practice that Heim (1993, 91–92) describes as a "working Platonism" that attempts to realize a long-held desire for an ideal sphere beyond the embodied, earthly ("contaminated" and "limiting") form.

The naturalization of technology is reflected in VR inventor and guru Jaron Lanier's (Biocca and Lanier 1992, 160) call for a "postsymbolic" communication. The cry for what amounts to "codeless communication" partakes of our need for extensibility yet blurs an understanding of this need with a particularly Western wish for transcendence. The desire for codeless communication or "direct" perception of external realities occurs within a cultural practice holding that technology is needed for extensibility and transcendence, even as the ambivalent relationship to machines identified by Carey and Quirk lingers on. This ambivalence surrounding technology reflects an uneasiness about communicating with others, as well as a recognition that in an American culture of extreme individualism, thought might not be private or fully autonomous after all but might also need to take place in public, might partly depend on language and learning, and therefore on others. Through technology's mediation, the individual can acknowledge his or her social and physical interdependency, if only temporarily, since the negatively conceived psychic impact of such an acknowledgment is given an emotional and spatial distance by the representational cast communications technologies demand. In seemingly paradoxical fashion, VEs as a new frontier for individual (self) discovery also achieve an appearance of "codeless" naturalness and direct iconographic "see-ability." In the old sublime, nature as a stage absorbed and humanized the machine. In the new sublime, the stage is technology that has absorbed nature and taken on the idealized components of nature's earlier ascribed qualities — decentralism, harmony, communion, peace, holism. One extends oneself via an immersive VE, hoping to experience technology's implicit promise of renewal by being absorbed into the technology. To extend the metaphor of the stage, if the old proscenium called attention to the space between audience and stage, the immersive nature of VEs collapses this space, merging the actors with the stage.

3. The Sensation of Ritual Space

As the critical histories of Virtual Reality and earlier technologies provided in the previous chapters indicate, the scientists, technicians, computer programmers, and hardware engineers who make possible VR and the creation of a variety of virtual environments — whether they be of military, educational, entertainment, commercial, or medical applications — operate within social contexts. VR at one level is a utilitarian technology, built because of, and responding to, human social needs and desires. It is a technological reproduction of the process of perceiving the real, yet that process is "filtered" through the social realities and embedded cultural assumptions of VR's creators and designers. VEs, however, partially because of their immersive nature, suggest their own ontological reality. The technology is premised on the creation of a "world" as "real" as the one we experience on a day-to-day mundane basis. Certainly this premise inveigles social acceptance and excitement surrounding VR. However, as indicated by the underlying language of the computer codes by which all VEs are constructed, the various virtual worlds VR makes possible are only in part based on the natural world. In many cases, they are largely, if not completely, socially constructed. This constitutes a significant portion of their cultural appeal: that users might play with identity freed from the constraints imposed by the natural world. As such, VR presupposes and assumes specific ways of looking at reality that are then built in to, and finally inflected by, the technology itself as part of a recursive or iterative process established between it and users. This chapter addresses some of these presuppositions and assumptions.

VR is often positioned as a communications device. Therefore, examining the presumptions it makes about the meaning and nature of communications is vital. In VR, an older understanding of communication as drawing people together in *ritual* gathering gets subsumed—though the technology still appeals to this understanding—under a more modern definition of communications as a *transmission* of messages across space. This change has important implications, which I discuss in this chapter. VEs also unsettle long-standing beliefs about the relationship among conception, perception, and sensation. How VR instantiates these relationships has direct bearing on concepts of self-identity, human connections to place, and an ongoing desire or yearning in the West for affective transcendence from the (embodied) "here and now." As suggested by the use of the term "Cartesian space" on the part of VR engineers in referring to the "world" on the other side of the interface, also central to the construction of virtual worlds are concepts of space, place, and landscape. VR is a spatial practice premised, in part, on a semi-explicit claim that it constitutes a meaningful place or places, and I analyze some of the underlying philosophical assumptions about space, place, and landscape designed into the technology.

Ritual Transmission

The introduction of the telegraph marks an important "rupture"—the "moment" when, for the first time, messages travel faster than the physical transportation vehicles, humans included, earlier necessary for message transmittal. The apocryphal curse "damn the messenger," implicitly pre-telegraphy, implies the consubstantiality of the individual carrying the message and the message itself and reflects a period when transportation and communication were fully one. We may now slam down the phone in disgust, but we no longer execute the instruments—the reluctant and winged "Mercurys" of old. A communication mechanism's form affects how meaningful content of messages is received.

In discussing the importance of telegraphy in reconstituting nineteenth-century decision making across a range of physical and social geographies, Carey (1983) notes that print technology cannot disseminate itself. Yet its speed of distribution was adequate for the territorial size of emerging European nation-states (Anderson 1991). The vast physical scale of the American state demanded greater instantaneity. The telegraph, introduced in 1844, initiated the separation between transportation and communications, as the information content of the technology, unlike print, could move from one place to another not only with minimal human intervention but also with greater speed than physical objects. As a communications device, the telegraph exemplifies the coming of symbols to control physical processes (Carey 1983, 304–5).

Because of the telegraph's speed, people came to equate communication with the transmission of messages across space. "The telegraph... allowed symbols to move independently of... transportation... [and] freed communication from the constraints of geography" (305). Although its practices have not entirely disappeared, an older emphasis on bounded ritual communication in place, with its forms of language and habitual social interactions, has gradually been displaced and marginalized (313). The telegraph allowed electrically communicated messages to be understood as operating differently from the transportation of people and material goods through space. More importantly, the telegraph allowed communication to control transportation—for example, the construction of telegraph lines parallel to rail lines allowed the coordinated scheduling of trains. Communication of messages superseded their transportation by humans, animals, and vehicles (see also Blondheim 1994).

In linking the United States informationally, the telegraph also affected what was considered newsworthy or interesting. Its cost meant that it carried only stories of national interest. A sense of community, previously localized and concrete, became imaginable nationally, albeit its being thinly spread across or linked by the wires. With almost every place having reliable access to more information, an earlier "city-state capitalism" gradually yielded to an emerging national commercial middle class linked to the telegraph and its "economy of the signal."[1]

VR, like the telegraph, is a communications technology, and VEs further complicate the relationship between ritual and communication. My understanding of experience in a VE as potentially a ritual of transmission complicates Carey's (1975; 1983) distinction between two contrasting understandings of communication within Western thought: (1) communications as transmission of information *through space*—a metaphor of spatial geography or transportation, and a practice that seeks to overcome the impediments of time and space. This view of communication has been gaining salience exponentially since the introduction of the telegraph. And (2) an older understanding of communication *as ritual*—the maintenance of society *in time* through representation of shared beliefs among people brought together in one place. Communication as ritual admits the sensed possibility for moral improvement through communicating. Communication rituals such as religious ceremonies or festivals, or secular gatherings such as a trial or an academic convocation, draw a group of people together in one place, often to bear witness to testimony of action. The place itself is a middle ground drawing together the disparate elements into communication. The first, now more prevalent meaning of communication associates it with the transmission and diffusion of goods and messages across space. VEs draw together these two meanings of communication to suggest, at a time when the "inbetween" meaning of place is perceived as in retreat, that the act of transmission itself becomes an ersatz place and constitutes a ritual act or performance.[2] On the one hand, as with communication as transmission through space, messages travel between geographically dispersed locales. On the other hand, VEs propose that this "space of distance" be modeled as a cyberspatial "room" where all might "gather together" in a manner that — at the scale of human bodies — *seems* to extend even as it dilutes or subverts the ritual component of communication.

Electronic technologies progressively undermine an older sense of place, one of the tasks of which was to undergird social abilities to perform copresent interactions with nearby people in a meaningful way (Meyrowitz 1985). An IT-saturated environment makes it increasingly difficult to argue that proximity might offer sufficient cause for interchanges between neighbors to be necessarily less representational or digitized than those conducted with people or machines situated across the planet. As the dark humor of figure 4 implies, psychic "distance" and insecurity are increasingly inserted by mediation into still-proximate spatial relationships when embodied, ritual forms of communication cede to transmissive communication across space. This insertion also depends on which conception of space is assumed. An absolute conception of space directs users toward transmissive forms of communications technologies — such as the telegraph — designed to send messages *across* an empty space that is conflated with distance as an impediment to be overcome. In contrast, ritual forms of communication understand space less as a distance whose effects can be altered by physical substances — though this is not denied — than as a possibility that grounds the basis for coming together. Ritual forms accord greater influence to the human intentionality and decision making that influences how people and things are spatially arranged in an interrelational fashion. Although it may at times seem experientially necessary to understand space as a distance to be overcome, and that thereby distance ought to be acknowledged and perhaps examined as having direct effects on how people gather, on its own the "space = distance" equation is arid and insufficient. It ignores the roles of human intentionality, agency, and influences instituted by different substances. Next-door neighbors might use the telephone to communicate, but this form of communication differs substantially from when they speak to one another from their respective front porches. Ar-

Figure 4. "Message Retrieval Disorder — The Neurosis of the Nineties." From *This Modern World*, copyright 15 November 1995 by Tom Tomorrow. Used by permission of the artist.

guably, the telephone conversation joins two individuals *across* a space-as-distance, whereas two embodied individuals addressing each other do so *within* a commonly shared space that they, in part, define. When mediation inserts a "psychic" distance, even among spatially proximate individuals, copresence is superseded by *telepresence*. A series of signs — verbal, written, or, in the case of VEs, iconographic — is intended to telegraph a meaning or concept across an empty space rather than within a space "peopled" with humans, objects, and things. The message or signal becomes the main event, and with VEs, an individual's representation becomes a signal available for transmission, extending and reformulating the dynamic first put into practice with the telegraph that both reifies and deifies the human as signal.

Imagine a scenario in which a group of international banking executives gather together for decision-making purposes in some form of virtually tele-embodied teleconferencing. The bankers may be dispersed geographically, but by donning head-mounted displays and separate exo-

VR Conferencing

This interesting application of telepresence combines video (taken at remote site) and VR technology to give users a very realistic method of holding meetings. Superior to existing video conferencing techniques, it could conceivably change today's reliance on costly, time-consuming business travel.

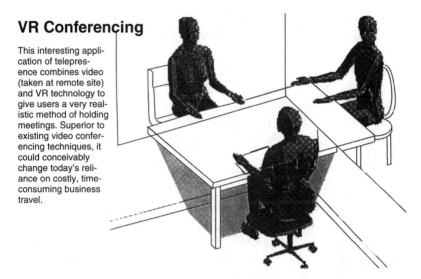

Figure 5. VR conferencing. Copyright 1994 Sun Microsystems, Inc. Copyright of this figure is owned by Sun Microsystems, Inc., and is used herein by permission.

skeletal body-tracking devices, they seem to be together in a virtual room. Major industry players, such as Sun Microsystems Computer Corporation, are promoting this "near future" in their literature, as figure 5 shows.

Although it is arguable that such scenarios may offer no material advantages over existing forms of communication, and therefore, though under design, might not be implemented, the VE proposed by Sun Microsystems suggests that VR will allow new forms of virtual communication. A computer would retain a programmed model of the virtual "room" in which the meeting is to occur and would transmit information among the sites where the bankers are located, in order to create the VE gathering space. As the bankers look around the room, the software and hardware of the programmed model and tracking device they each wear generate continually updated images of the room and occupants that are geometrically accurate for the viewing position of each user. Ritual communication would take place within an immersive framework or landscape incorporating representational aspects, or "avatars," of the users' imaginative selves while leaving their bodies "behind." In such a "gathering" intended to facilitate economic decision making, users might not elect, or have the choice, to "don" imaginative iconic facial identities. A separate video bandwidth dedicated to transmitting facial expression seems likely, as is the case in HITL's and Fujitsu's Green-Space prototype. Photographically true facial representations are merged into idealized backgrounds. The bankers may seem to meet in the most sumptuous of meeting rooms, but it is unlikely they would be permitted to appear, for example, as thunderstorms if angry.

Communications theory often construes space as a psychic or material impediment or distance to be bridged. It is arguably the case that speech itself—the voiced words that communications technologies represent, for example in book form, or transmitted via radio broadcast—is a representation of the thought it attempts to express. Speech extends the speaker's range beyond his or her corporeal limits. However, contemporary considerations of the power of language often fail to consider that at the scale where direct human speech as a communication retains affect, language as voice is part of the body. Spoken language is Janus-faced, pointing "back" to listeners-receivers in the direction of the speaker-sender's body, yet also "forward" toward human extensibility across space. Communications technologies such as the telephone that transmit human speech as a message across space omit to some extent the ritual components of language, particularly those that are embodied. Although telephone use has generated its own kinds of rituals—the prefatory "hello," for example, or "speaking" when the speaker is requested, telephony abets forgetting the meaningful role of the body in copresent forms of language interaction. Gestural communication, for example, does not transmit via telephone calls; nor do the facial movements that are a part of embodied communication. Communication technologies further develop a specific and direct relationship between disembodied *representation* and the *distance* a representation or symbol must cross.[3] Although cultural elites have long benefited from their ability to send messages across considerable distances, it is arguable that for the majority of, say, an illiterate medieval pre-European peasantry, communication meant copresence between individuals or groups.[4] The increasing power of communications technologies has permitted a partial democratization of the power of communication across distance. Yet with this has arisen the privileging of representation at the expense of human bodies. With respect to text-based communications technologies, this is indicated in the widely held belief that freedom of the press and freedom of speech are the same (see Innis 1951). The press, and the text by which it communicates, facilitates the making over of an earlier more embodied communicatory practice into one more approximating "pure" information, any bias notwithstanding. The commodification of the speech act via text is placed on an equal, some would say greater, footing than the referent—speech—to which it refers. In a similar manner, the Sun Microsystems variation on embodied telepresence privileges representation of the bankers at the expense of their bodies, even as it *seems* to acknowledge the importance of bodies in ritual communication. The issue of seeming is important here, and a brief excursus linking seeming as a specific form of representation to embodiment, social relations, and VEs is germane.

Theorizing bodies as emblems—and I am suggesting that this is the effect of Sun Microsystems' proposal for, and depiction of, a virtual teleconferencing environment—calls to mind connections between a medieval sense of understanding that underpins the affect, or transcendent potential, of emblems, and the power that lies in the contemporary implosion of image, reality, and discourse identified as simulation and made visible in virtual technologies. The body electronic, made "present" through electronic space as a picture-image with or without supporting spoken text, can be argued to be such an emblem with or without a firm connection to its referent. Writing of former British prime minister Margaret Thatcher's political power and its connection to TV, John Fiske notes that her "political power is the *same* as her image power, her power to *do* is the same as her power to *seem*" (1991, 57). I am assuming here that Fiske assumes her image most often to have had audio accompaniment. Baroque emblematics almost always employed a picture-image, most often incorporating the human body and/or an architectural motif in conjunction with a subtext included to guide and police the viewer-reader's interpretation (Buck-Morss 1989). The English word "emblem" is quite close to "resemble," but closer perhaps to the French *sembler*—to seem. When it is recalled that the power of the pre-Enlightened embodied monarch depended on the *seemliness* of his physical representation in public—itself an act of becoming—the greater metaphysical affect, or apparent transcendent potential, of seeming rather than merely appearing is underscored.

Seeming is explicitly included in the theorization of virtuality offered by Ted Nelson, a key player in, and critic of, virtual technologies research since the early 1960s.

> An interactive computer system is a series of presentations intended to affect the mind in a certain way, just like a movie. This is not a casual analogy; this is the central issue.
> I use the term "virtual" in its traditional sense, an opposite of "real." The *reality* of a movie includes how the scenery was painted and where

the actors were repositioned between shots, but who cares? The *virtuality* of the movie is *what seems to be in it.* The *reality* of an interactive system includes its data structure and what language it's programmed in—but again, who cares? The important concern is, *what does it seem to be?*

A "virtuality," then, is a structure of seeming—the conceptual feel of what is created.... It is this environment, and its response qualities and feel, that matter—not the irrelevant "reality" of implementation details. And to create this seeming, as an integrated whole, is the true task of designing and implementing the virtuality. (Cited in Rheingold 1991, 177)

The issue of *seeming* provides links between virtual technologies and the electronic communications devices that precede them, yet it also provides a means of theorizing how the form of virtual technologies affects the relationship between conception and perception—the subject of the next section—in novel ways.

At least since Saussure's theory of signs, it has been accepted that words or linguistic signs are two-sided. They consist of a signifier (the sound if spoken) and a signified (the meaning to which the signifier refers). The two-sidedness of words creates a distance or a gap between the representational status of the word-as-communication and that to which it refers. This severing of material and conceptual elements, of body from idea, flowing from the need to use representational devices to transmit information across space understood as distance, inserts a difficult-to-avoid metaphysics into what it is to communicate. The severing supports belief that communication might remove the need for material locations where embodied relationality or community *take place.* Community—admittedly a freighted term—is in part how and through which we are defined by our experiences of others. Our existence is confirmed and made meaningful because, on some level, we are given to each other. This agency is foundational in making meaning possible (Nancy 1991). In the absence of embodied rituals wherein meaning is produced, maintained, and repaired, communication (the text of becoming) may overtake existence (the voice of being), and agency then derides, among other things, repose and the time that contemplation demands. Enter sophisticated iconographic communications devices as forms of "metaphysical technology," or magical empiricism, intended to heal, but only virtually, the distress that flows from the psychic distance between representation and meaning, and the experiential distance between discrete latter-day monads responsible for producing their own meaning and all that lies external to themselves. To achieve transcendence, people flirt with becoming information. If achieved, this would be *ex stasis*—out of the human body's stance but less into the arms of such possibilities as love and sex that bodies, at least in part, make possible and locate, than into the visibilized world of language, fiber optics, and trance.

Conception, Perception, and Sensation

Virtual technologies unsettle existing relationships among the roles of *conception, perception,* and *sensation.* The design of immersive virtual technologies, the implicit desire to view them as "transcendence machines," and the drive to blur virtuality and reality in military and commercial applications, work synergistically to collapse distinctions between the conceptions built in to VEs and the perceptive faculties of users.

Conception implies the act or power of forming a notion or an idea. A conception might be a plan or *sketch of something not actually existing,* as in an "artist's conception." "Concept" is an idea of something formed by mentally combining all of its characteristics or particulars, or a directly conceived or intuited object of thought—an idea. "Con-cept" blends an understanding of "together" with "seizing or taming."

Dictionaries distinguish conception from perception and sensation. "Perception" can mean an act or faculty of apprehending by means of the senses or of the mind. A more psychologized definition implies a single unified awareness derived from sensory processes while a stimulus (a book, for example) is present. "Perception's" Middle English root suggests "a taking in" or a gathering. Agency is present, but perception is more receptive than what is contained in "conception's" original meaning of "to seize or tame."

Sensation derives from the Latin *sentire,* meaning to perceive or to feel. More so than perception, it implies sense stimulation. The meaning of "perception" straddles conception and sensation (Reese 1980, 100). Sensation, with its closer ties to external stimulus than is the case for perception, offers the better contrast to conception. Reese's proposal bears some similarity to David Hume's understanding of the difference between ideas and impressions. For Hume, experience is composed of ideas and impressions. Impressions have the greater force and vividness. Although ideas can be traced to impressions, the latter cannot be traced to any source. Like sensation, impressions arise from the external world, though proof of this is not available to us through experience.

VR's promise of interactivity is based on a twofold process. The iconographic virtual worlds represent the conceptions of military, commercial, and scientific interests along with those of the software designers who interpret these conceptions and then write the programs that, along with the hardware, translate these interests into VEs. The technology's immersive quality then combines with its vivid visual imagery to give the impression that it offers an experience of unmediated sensation (referred to by VR designers as "direct perception") when in fact it presents a highly mediated series of conceptions or ideas.

Although interactivity anticipates a limited range of user input, VEs actually propose that users experience someone else's conceptions (in itself not a new phenomenon) represented as highly vivid sensations experienced via a process of immersion (a new phenomenon that reduces the distance or space between object and subject). This works to confuse users' perceptual experience with the conceptions programmed and designed into the machine. "Everything in the field of view is presented to the senses. . . . VR is a literal enactment of Cartesian ontology, *cocooning a person as an isolated subject within a field of sensations* and claiming that everything is there, presented to the subject" (Coyne 1994, 68; emphasis added). Hence Jaron Lanier's (1992) somewhat spurious suggestion that in VEs we will communicate without codes. In other words, VR will present a "postsymbolic" visual language of such excellent technical refinement that, like subtle artworks of great power, VEs will pass for or merge seamlessly with perception itself.[5]

When I read a novel, I might experience the sensation of being somewhere else. What the novel proposes to me is in the nature of an "artist's conception." I fill in the details by extrapolating vivid sensations from the novel's more abstract conceptions, and this agency is part of my engagement with the text. Virtual technology is being developed around a view of perception as a passive conduit for transmitting sense data between the discrete exterior world and the self "within." Theories of perception inform theories of communication, and a passive view of perception translates into a theory of communication holding that messages pass in a direct and unmediated fashion between and among people and things along channels theorized as passive conduits.

With a VE, the machine "thinks" the form of the image. Unlike how a novel is experienced, in a VE, the mental work required to extrapolate sensation seems unnecessary, as a central promise of this technology is sensation itself. The nature of immersion is to make users feel drenched in sensation. The images have the power to arrest or seize the viewer's perceptive faculties. I would note that the "external stimulus" of this sensation is entirely cultural. The natural world is not in view. At the scale of the individual user, as sensation and perception become detached from each other, the genesis and activity of vivid sensation — and by extension conception, ideation, and critical reflection — are relocated from the perceiver to the technology and its inventors and authors.

Many VR scientists and engineers argue that the more accurate and numerous the data, the more realistic the representation. This differs from the novel form, which, in successful examples, conveys information in such a manner so that a part of a reader's experience is creative engagement with the text, leading to ideation or production of a mental "construct." If this second, more productive understanding of perception were applied to virtual technology development, it might suggest that the technology would "not have to strive for realism through better and more accurate sensory input" (Coyne 1994, 66). When perception is understood as a passive conduit, its role within corporeal intelligence is minimized. With VE design, this has led to a naturalized assumption that the communicatory relationships among machine, person, conception, sensation, and perception position the machine as the primary sender. Therefore, sensation is brought into the conceptual orbit of the technology, leaving the tamed viewer with perception stripped of its active meanings.

These concerns relate to a conception of space suggested by VEs operating in this manner. In the room in which I am seated as I write, when I turn my eyes to look around me, my embodied vision perceives different aspects of the environment. I see the monitor on which these typed words are displayed. Turning my head slightly, I see the bookshelf, then the wall-mounted heating unit, the stain on the carpet, et cetera. If I step back through the archway linking this room with the adjacent one, I can see these "discrete objects in space" in one fell swoop. In a VE such as the virtual kitchen offered by the Matsushita Corporation, when I turn my head to look around the virtual space, the space reconfigures before my eyes. Space itself may seem to have magical agency in the virtual world. Virtual space updates itself depending on how the technology reads my own body's movement. In the virtual military simulation discussed in the introduction, the computer reads me like an "other" in relation to its own perceptual abilities, so that it can then offer me a "perspective" from which my perception will be saturated with its imagery. In this way, virtual technology achieves a new form of spatialized power, based on unseen computational abilities with which my body is rendered complicit. In real life, to the eye, objects are arrayed in space, and I perceive this array differently as I move my embodied vision among them. In VEs, I depend on "the kindness of strangers," as it were, who in their programming and software designs first conceive and represent this array to me in order that I might, after this fact, perceive its simulation. As a process, conception precedes perception here in contrast to embodied reality, where perception of space precedes any ability on my part to give concept or meaning to my initial perception. The book also performs such a task but does not collapse the actual space between itself and its reader in the same wraparound fashion as do VEs. In VEs I am perceiving a "space," the "building blocks" of which are other peoples' encoded conceptions or uses of language. This suggests to me the loss of my perception's primacy, and by extension, loss of self-reflexive abilities supported, for example, by the more difficult decoding process abstract print demands. Audiovisual perception is important in VEs (and to a much lesser degree so is touch), but these become secondary and more pacified. Partly this is so because I need move my body very little in VEs to perceive change and spatial movement. And partly this is so because in real life my experience of space takes place primarily in my perceiving it. In VEs, I assent to a kind of double recursivity—the world designed *by* humans so that it might seem, if only as a second-order vision, to be designed *for* humans. A "world designed for man" seems very similar to the "world as the condition of the subject" that Elizabeth Grosz (1994, 96) identifies in her discussion of Merleau-Ponty and his phenomenological approach to body meaning. In their interactivity, VEs seem to offer the world as just such a condition. But this offer depends on hubris. It is conditional on users accepting as a moral good a reduction of the sensory interplay between people and their lived worlds to a concept of "world picture" from which the nonhuman natural world has been excluded.

Spaces, Places, and Landscapes

The previous sections of this chapter have implicitly discussed conceptions of space as part of the virtual world on the "other side" of the computer interface on which VR relies. Whether the theorist is Aristotle, Euclid, Newton, or Descartes, space is theorized in different ways. Certain aspects of each thinker's theory, however, are incorporated and conflated in VEs. In attempting to shed light on historically contingent, complex, and nested sets of meanings and interpretations of space, the account of spatial concepts that follows is organized chronologically so that the reader may gain a better sense not only of how space has been conceptualized but also why this matters to the theorization, construction, occupation of, and resistance to virtual "worlds." In this section, I discuss these philosophical underpinnings, as well as the related ideas of place and landscape on which VR in part depends.

Spaces

> If two different authors use the words "red," "hard," or "disappointed," no one doubts that they mean approximately the same thing, because these words are connected with elementary experiences in a manner which is difficult to misinterpret. But in the case of words such as "place" or "space,"... there exists a far-reaching uncertainty of interpretation. (Einstein 1969, xii)

Albert Einstein notes the complexity that attends theorizing the space and places of the real world. His comment serves as a caveat to my own requirement to distinguish among different concepts of space and place, and where each of these may lead with respect to the meaning of VEs. The discussion in this section thins out a dense set of ideas about space that is more fully examined at various points in subsequent chapters. For example, to consider theories of space is explicitly to deal with representations in language — an issue pursued more fully in the discussion of metaphor in chapter 5.

The theorization of space has a contentious history and practice, and spatial concepts remain subject to ongoing debate and reformulation. VR adds to this ferment. Its developers explicitly call on spatial terms. "Inside/outside," "world," "cyberspace," "theater," "gradient," "room," "platform," and "Cartesian space" are terms used to facilitate both public and personal conceptualization of this new technology and social practice. A VE is a representational space that relies on absolute and relational concepts of space. Immersive virtual technologies are constructions — visual metaphors or representations of absolute space. As such, they promote a long-standing and contested belief that space exerts an independent force. However, a VE also subverts what might be called the hierarchy of scale implicitly associated with these concepts, thereby also suggesting that the "structure" of space might not be as constant or universal as absolute space traditionally might be taken to imply. Absolute space suggests macrolevel or "big picture" realities. Experientially, relative space accords more closely with individual meaning, and relational space may suggest an ability to imagine a continuum or at least linkages between the meanings of absolute and relative space. Although VEs are based on Euclidean geometry and a Cartesian grid of absolute space, users will have some relative ability to manipulate how space (along with distance and motion) and objects are represented and relate to one another "therein."

In line with Aristotle's belief in the impossibility of a vacuum, he identified space with place (Jammer 1969, 53). Aristotle held that everything that exists is in a place and could not be so without place.[6] Metaphysically, all things are in space, but space is never *in* any other thing. Space is not the same as matter for Aristotle, though it is a continuous quantity, has magnitude, and can be measured. Space establishes limits to the bodies in it, preventing these bodies from becoming "indefinitely" small or large (10). The place of a thing is not a part of the thing itself, but what contains it. The place of a thing and the thing itself have the same size, but a thing can leave the place where it is and move to another place. From these findings, Aristotle offered a definition of place as "the adjacent [or inner] boundary of the containing body... [which is]... everywhere in contact with the contained" (20). Within this *relational* understanding, space synthesizes; it becomes the sum total of all the places of the world or universe occupied by bodies (11). However, Aristotle also understood space and time as categories that facilitate naming and classification of sense evidence. The status of Aristotelian space as a concept is unclear, and his was really a theory of place or one of how bodies are positioned in space, rather than a theory of space per se (15).[7]

Early-modern thinkers conceiving of space inherited contradictory notions from the ancient Greeks.[8] Descartes defined matter as infinite extension, and in his theorizations, the whole of space is filled with matter and cannot be empty.[9] Positing matter as extension, Descartes postulated that motion was possible due to the "subtlety" of the matter that fills all space (Reese 1980, 543). He believed no less than Newton that space was absolute, though he conceived of it in a different way. Cartesian space is extensible without limit and becomes absolute as an object that dominates, by containing all senses and all bodies (Lefebvre 1992, 2). Cartesian space "consists simply of the relations among extended objects" (Curry 1996, 10). For both Descartes and Newton, the abstract, isotropic, infinite, and absolute representational space of Euclidean geometry—itself a description or model of space and spatial relations, and a synthesis of earlier and competing theorems of geometry—was available for mapping over the geographic places of the real world. Alexandre Koyré has commented that the modern replacement of an Aristotelian conception of space—one that understood the world as a closed and "differentiated set of innerworldly places" (1957, viii)—with that of Euclidean geometry led to an understanding of real space[10] as identical with the infinite, open, and homogeneous extension that this geometry postulated.[11]

Newton conceived of absolute space[12] as God's sensorium.[13] He defined place as "a part of space which a body takes up" (Newton 1946, 6). Although Newton here refers to "parts" of space, Koyré (1957, 162, 165) suggests that Newtonian absolute space is indivisible and an entity in and of itself. Its abstract division may be conceptualized and represented; however, space as an indivisible entity permits introduction of the notion of a vacuum (or void) between objects in Newtonian absolute space.[14]

In contrast to Descartes — who held that all of space is filled through motion on the part of "corpuscular" bodies — Newtonian absolute space is empty, and distance between objects has an effect.[15] It is possible to see here an earlier form of the argument equating absolute space and distance. Although we do not generally conflate space with distance, we understand the former's influence in terms of the latter. I would note, however, that whereas an existential reality precedes or is before us, distance becomes phenomenologically real when we move through it. That we move our bodies through space somewhat relaxes distinctions between separate understandings of distance and movement, since to move is to understand the meaning of space, as well as to "cover ground."

Einstein (1969) explains the core idea of Newtonian absolute space with the following example. Imagine a cardboard box. The space within this box is not the same as the matter that constitutes the box, and one can extend this notion to arrive at what Einstein refers to as "independent (absolute) space, unlimited in extent, and in which all material objects are contained" (xiii). This conception of space achieves a meaning that can be freed from any connection with specific material objects. A material object not situated in such a space is inconceivable, yet within the parameters of Newtonian absolute space, an empty space is plausible. "Space then appears as a reality which in a certain sense is superior to the material world" (xii). Cartesian absolute space is predicated more on a "positional quality" of material objects: without material objects, such a space is inconceivable (xiii). The difference in conceptual inflection between an absolute Cartesian space conceived within the relationships among the positionality of objects, and the movement of objects within empty absolute Newtonian space, Einstein argues, has been somewhat reconciled by recourse to the system of Cartesian coordinates (xiii).

Roger Jones (1982, 17) observes that contemporary science's assumption of a "separate, quantifiable, objective world" has led to a concept of absolute space based on operational definitions and physical measure, along with a metaphysics according this space independent agency. Such a quantifiable and objective world is the essential premise of Euclidean geometry, and it is Euclid who first provides such "tools" for thinking about space and spatial relations in this fashion. Writing as a physicist,

Jones argues that verifiable measure matters little to "the essence and experience of space itself—the basic mystery in which we all participate, which permeates our every act and thought, and whose ubiquitous presence we accept unconsciously as synonymous with our very existence in the world" (17).[16]

As the concepts of absolute and relative space have come to be understood within physics, absolute space does not precede objects. Neither do objects create or "adjust" space by virtue of their relationships with one another. Space remains separate from the existence and distribution of matter. However, in the "arena" of popular understanding—and the community of virtual researchers—these distinctions intermingle with the legacy of Aristotle's notion of space/place as *preceding* that which is located within it. This is reflected in the continuing, if poorly articulated, belief exemplified in a VE that space precedes all things, which come to be contained within its sphere. If space is believed to precede objects, it is also easier to believe it has a power akin to "first cause."

Absolute space, with its rigid structure, is believed to have independent agency, exert physical effects, yet remain immune to influence (Sack 1980, 55). It provides a framework for the organization of social relations (the subtending function absolute space provides for relative space, as noted by Koyré [see note 12]). Relational space, like relative space, can be acted upon. However, like absolute space, relational space offers a matrix for the positioning of objects, though it exerts no physical effects on its own (55).

In relational space, things exist by virtue of their interdependency with other processes and things. This accords a certain privilege to distance per se—the existential reality that is not affected by the placing of things, say, on an axis running between *a* and *b*. However, the effect of the "space of distance" represented or established by this axis is influenced by the placing of things into this distance. Relational space also bears strong similarity to a concept of space proposed by Newton's contemporary, Gottfried Leibniz. He theorized space as a *system* of relations (Jammer 1969, 118), or the "order of possible co-existences" (Reese 1980, 299). Leibniz asserted that space derives from interrelations of the things that together compose the universe (543),[17] and it is not an independent thing apart from material objects (Einstein 1969, xiv). This privileging of interrelations between things also implies a privileging of movement across space, and the potential to conceive this movement as being one and the same as space itself. When space results from movement, places become objects in a space of continuous circulation — and the more synthetic and relational Aristotelian concept of place is inverted.

VEs, then, merge absolute and relational concepts of space. The absolute grid of Euclidean geometry that influenced both Descartes's and Newton's concepts of space becomes an "originary" Cartesian space[18] on which spatial interactions among agents, and agents and things, are visually represented in a relational fashion. This humanly created representational space is made an agent in and of itself. Yet this understanding and application of absolute space as the grid or "stage" of performance is melded to a spatial relativity that operates at the scale of the different actors, their interactions among one another, and even among spatialized "parts" of the same actor. The "Cartesian space" of a VE identified, for example, by NASA researchers Stephen Ellis (1991a; 1991b) and Mary Kaiser (1991) offers infinite extension across an absolute grid that situates an array of digitized representations. Yet conceptually, though this is a created space constituted in matter, at the moment of its original fabrication, this space is a conceptual void, like a blank sheet of paper awaiting the placing (or writing) of things. Any eventual user will grasp this space only through interacting with objects "within it." Although the gridded space of a VE is "absolute," Ellis (1991b), for example, discusses communication within it in a manner that seems analogous to distance. He notes that communication in a VE is interactive in at least two directions: the picture or spatial display has effects on the viewer, and the viewer influences the picture.[19] Koyré (1957, 114) offers a clue to understanding the conflation of space and objects that come together in a VE when he argues that the "Cartesian geometricization of being" destroys the distinction between space and the things that are in space.

This necessarily selective account of concepts of space suggests that within Western thought at least, how space is represented, described, theorized, and conceptualized in language and image is as important as how it is experienced. An experience of space either before or apart from an effort to communicate this experience leaves one mute in the face of the demands of social relations. One must be able to represent the experience to others in a more or less sensical fashion. Yet I am interested in how an experience of space — influenced by spatial concepts conceived within cultural contexts and social relations — in itself influences what it is to communicate. Henri Lefebvre (1992, 12) captures something of this conundrum that attends attempting to describe ontological reality.

Space considered in isolation is an empty abstraction; likewise energy and time. Although in one sense this "substance" is hard to conceive of, most of all at the cosmic level, it is also true to say that evidence of its existence stares us in the face: our senses and our thoughts apprehend nothing else.

Yet how is space to be conceived today? Is it, for example, to be conceptualized along the lines of a scale of integration; or is it a kind of ecological matrix or ground that integrates objects; or is it a way simultaneously to interrelate the kinds of integrations that are the essence of places (see Nunes 1991, 16)? Theorists can debate spatial concepts (and spatial metaphors) and argue about space's constitution, but is space so "fundamental" that it can elude language, which it precedes (yet *we* require for communicating its meaning), while also forming part of meaning's constitution? Sack (1992, 19) finds that space "is . . . an ontological category in nature . . . fundamental to agency . . . and constitutes the experience of being in the world." Social relations, agency, and meanings take place in, and are influenced by, space and how it is used. James J. Gibson (1966) — whose theories of vision have been so influential in VE research — offers an understanding of space that falls short of a "definition," yet to which I am somewhat partial, as it suggests an "ontological agency" that nevertheless remains open to interpretation and a recognition that humans confer meaning on space. In "The Air as a Medium," he writes,

> The atmosphere . . . is a medium. A medium permits more or less unhindered movements of animals and displacements of objects. Fundamentally, I suggest, this is what is meant by "space." But a medium . . . also permits the *flow of information*. It permits the flux of light, it transmits vibration, and it mediates the diffusion of volatile substances. (14)

The social sciences receive a view of space, "the characteristics of which are beyond their purview to explain" (Sack 1980, 57). Physics has been successful in arguing that the Euclidean framework of space is "immune from the effects of human behaviour" largely owing to our activities being of an insignificant energy and mass to affect the geometry of space (57). Roger Jones argues that space is one of the deepest "expressions" of our consciousness (1982, 50). Most attempts to define space are prescriptions "for assigning a measure or number to space in a precise and reproducible way" (17) — an activity that, as Koyré (1957) notes, permits space to be abstractly divided and therefore represented.

Although space is distinguishable from the measure imposed upon it, and the language used in its conceptualization, experiences of locating the self and others gain meaning and coherence through spatial representation; otherwise, such phenomena as VEs could have no cultural point of purchase. People define space, and it defines them (Ardener 1981, 11–13); and the development of human perception is wedded to how space is experienced (Tuan 1977). We see ourselves located as points in space, separated by spatial distance from others, and connected by light and sound that travel across space. As a metaphor, space "organizes and gives meaning to our tangled, amalgamated experiences through its attributes of place and distance" (R. Jones 1982, 50–52).[20]

Benedikt (1992b, 125), writing about VR, arrives at a similar determination. "Space and time . . . appear to constitute a level of reality below which no more fundamental layers can be discerned . . . a universal attribute of Being that cannot be done away with." Benedikt is operating within a Kantian framework. For Kant — who revives and renovates the Aristotelian categories of space and time as ontological qualities — space is a priori to experience, a transcendental understanding that nonetheless allows for a temporal relation between space and objects. The Kantian *phenomenon,* positioned between human understanding and an a priori framework of space and time, may suggest that for all practical purposes, what exists is only what our consciousness can know. A variation of this latter stance is adopted by George Berkeley in his theory of perception and sensed experience (see chapter 4). However, Kantian space assumes Newtonian absolute space as a physical reality, though space is also asserted to be an a priori condition of experience not derived from this experience and not attached to matter. Kantian space is not an object of perception but a mode of perceiving objects (Jammer 1969, 138). This transcendental aspect of Kantian space as a condition for the possibility of experiencing the world has never achieved the widespread cultural naturalization accorded to his (and Newton's) concept of absolute space. However, the phenomenon has been taken by some as offering support for asserting that existence — and, by extrapolation, what we identify as space — might be secondary to human consciousness and social relations (see, for example, Sass 1994, 90). Yet I would maintain that even if a knowing consciousness thinks itself paramount, it still understands space as one of its deepest expressions. Kant's main point here is that space and time are a priori necessary conditions of sense-experience (Copleston 1994, 238). The thing-in-itself exists whether or not it is an object of experience. I would note that when existence is made secondary to consciousness, so too has it become secondary to communication and its technologies, for the immateriality of consciousness implies an insecurity vis-à-vis the material world. Existence and space get reconceptualized as pure functions of communications, and the natural world seems swallowed by wholly representable social relations.[21]

Curry (1996) comments on the difficult issues raised by competing yet interpenetrating concepts of space. His observations are useful in thinking through the complexity inherent in conceptualizing space with the ways that immersive virtual technologies, embedded within cultural contexts, synthesize concepts of space often held to be discrete from one another. For Curry, when we conceive of the world as a place and when we see the world as a set of places operating in a variety of hierarchical ways, we are latter-day Aristotelians (8). Yet in imagining the world as an infinite array or geometry capable of being parsed into a featureless grid, we owe a debt to Euclid and Newton. Curry suggests that a pivotal component of the enduring cultural saliency of Newtonian space is its power and clarity as an *image or world picture* (5), which is supported in numerous ways by the modern project (8). Moreover, geographers understand that space is necessary psychologically in offering humans something against which we might stand out as knowing agents. Space is a condition of (and a priori to) our understanding the world. Such an assertion is neo-Kantian. We are also neo-Kantians when we relativize concepts of space according to aesthetics, morality, economics, and so forth. So too when we acknowledge the ways by which, for example, different cultures organize space differently by virtue of how they array objects "within" it. The form of space can vary. Different concepts have been developed that are appropriate to the disciplinary contexts within which they will have been used (24).

The spatial technologies of virtual worlds operate in a similar amalgamating yet parsing fashion. They bring all these understandings together under one umbrella. The difficulty lies in the fact that the absolute aspects of VEs are occluded by the suggestion and demand that each user's agency, pleasure, or duty in part will reside in defining and viewing the umbrella in radically different ways. One might say that an image of cyberspace as an absolute and infinitely extensible representational world is blurred by privileging what seems akin to the user's hyperpersonalized "mental map" of this world. Although a VE is based on a gridded Cartesian space, and is Euclidean in that it confirms a belief that no space need exist "outside of that created within the demonstration itself" (9), theorists and designers also suggest that users will expand, shrink, or otherwise reshape this space in varying ways, thereby conferring at an individual scale an aspect of relative space onto the macroscale absolute concept.

Such a fragmenting and relativistic dynamic, moreover, is hostile to ways by which the meanings of places are negotiated through shared experience. These experiences operate along a scale running from the psychically integrated individual's ability to share the meaning of a single experience among its various components of self, to the ability of a group to engage in meaningful communication as a result of its more or less having come to terms with or negotiated the meaning of a shared experience.

Place and Places

Place — an archaic notion for some, but whose ongoing saliency I wish to support — can be conceptualized as drawing together (as does human consciousness, which could not exist without place) the spheres of nature, meaning, and social relations that modern conceptualizations pull apart for analysis. Place offers a means to center phenomena. Although I could scarcely disagree, for example, with Einstein's materialist definition of place as "first of all a (small) portion of the earth's surface identified by a name" (1969, xiii), I am more interested in the experiential meaning of place. An experience of being in actual places suggests something of the agency that resides in phenomena. Places are not arithmetically apportioned. They are not made up of neat thirds of meaning, nature, and social relations with the agent always somehow apart and looking "on" or "into" them as discrete abstractions. Any attempt to link meaning, nature, and social relations additively would be labeled in error by those asserting that Being had been forgotten, and by those claiming that the sphere of Being includes beings or agents.

Each place is uniquely constituted. Imagining a series of intersecting circles in a Venn diagram as the interplay of a number of different places gives a hint of the complexity and interpenetration of places and peoples in the world. In addition, it may suggest why places and their conceptualization need not be understood as reducible to the old chorography of discrete particularisms once favored by geography. This is, however, also why places are "messy" for nomothetic social science approaches.

Humans are always in places, which are at least partly composed of, and created by, the nonhuman physical world. I understand places and people as constituting interdependent, fluid, and relational unities, ones with very leaky and imprecise borders. Indeed, part of the function of the place naming to which Einstein refers is to confer a cultural stability, through naming, atop this unavoidable imprecision. Geographic places and how we think about them are fundamental to how "we make sense of the world and *through* which we act" (Sack 1992, 10; emphasis added). I italicize *through* to draw attention to the relationship between the places within which we dwell and the grounding of any ethical potential for human agency. Victor Walter writes of places as being "expressive intelligibility" (1988, 2) apprehended through sensuous reasoning (121), and I want to link his succinct phrase to a suggestion that how we sense place is not so different from places themselves. We give places meaning, and in return they offer us existential support. At the same time, we literally make the places we inhabit "with sticks and stones. A built object organizes space, transforming it into place" (Tuan 1980, 6).

The *concept* of place is itself a classifying and organizing principle, with places having continued to exist across all the "epochs" identified by the use of calendar time. Those who would write off both the concept of place and the role of concrete places in the world as overly traditional, limiting of human freedoms, or even the bulwark of a reactionary politics, tend to assert that identity formation is anchored solely in social relations or the flow of history, with its "relentless march" through and over the places of the world. Social constructionist arguments tend to deride places for being constraints on freedom rather than locations for the latter's performativity. Hence the view—which John Agnew (1989, 9–10) links to a privileging of concepts of community and class—that place is static and only serves the forces of reaction (see de Certeau 1984; Poster 1990). At the same time, however, the academy launches a sustained search for the missing human body and for links between the "discrete" spheres of nature and society within what has become a technologized dimension more of communication between signs or traces than people themselves.

For example, Michel de Certeau writes of space as a practiced place, of place as an order "in accord with which elements are distributed in relationships of coexistence" (1984, 117). For de Certeau, places are loci of control, and the strategies of common people are spatial. It is not so much that these people are incapable of planning as that they are continually denied an official history, which is arrogated to official texts and "places." Common people must continually reclaim shards of territory or "space,"[22] from which they organize cultural and resistance activities. Although the history of common people is suppressed or denied, the spatial reality of their laboring bodies cannot be taken away from them, and it is this reality that organizes a spatial politics, however temporary and partial.

Much is written about traditional peoples and marginalized groups as having no "history." In a modern sense, this may be true. Instead, however, these people have places within which they are constituted as embodied beings, and that engage with other places in more or less fluid or codified relationships. Place serves as an adequate matrix for organizing meaning. The people, the things, and their places, memory included, all retain ontological status.

Places — marketplaces, universities, family rooms, nightclubs, malls, cathedrals — make exchange between people, and between people and things, possible. This exchange helps provide us with an identity in equal measure to the "flow" of time that runs through all places in the modern period. Exchange is what gives places their potential for change, even as a place itself is destroyed, altered, endures or continues. This exchange, along with the degree of engagement entailed, can be difficult to represent precisely, and with the widespread belief in symbolic representation as the best means for accessing truth, a messy process of exchange is subject to accusations of irrelevance in the face of seemingly more powerful devices such as "structure," "framework," "system," or even the more nuanced "network." Places are the middle ground where all the different human and nonhuman elements come together, and this was perhaps Richard Hartshorne's (1959) greatest insight, reflected in his notion of "element complexes" and their drawing together of human and nonhuman "geographic individuals" of discrete and continuous natures. Nicholas Entrikin's (1991) identification of the "inbetweenness" of place also captures something of the meaning this middle ground offers to its "users."[23] Such users, elements, or beings have many different and overlapping dates of birth and death, origin and expiry, but their relationality within places also serves to distinguish them from one another, even as they interdepend spatially and temporally.

Our *sense of place* is memory qualified and deepened through imagination. Memory and imagination depend on experiences and take place in our bodies, which act as sensory mediators of, and witnesses to, this experience. There is a unity between bodily sense and place, but it can be broken if the flesh is set aside, for example, when we communicate across distances via representations. Setting aside recognition of our bodies' importance, as does *cogito ergo sum,* not only overemphasizes a distinction between how we sense a place and any subsequent conception of place. It also allows places to be viewed as containers or stages for human performativity, or worse, as "sites" for resource extraction, because our core of self is alienated from the myriad materialities and spirits of places. Such an alienation leads us to see places as intermediary devices, or mere "sites" or "spatialities" through which we might pass, but of which we are never fully part.

Place "is an organized world of meaning," and its identity comes from "dramatizing the aspirations, needs, and functional rhythms of personal and group life" (Tuan 1977, 179, 178). Place is also a "field of care" that is qualitative, elastic, flexible, and sympathetic as need demands (Tuan 1982, 155). The need for novelty, and the desire for transcendence from the here and now, are expressed in the conceptual and actual extending of oneself beyond the immediate place where one is. However, the endless disembodied extensibility across space that proponents of VR anticipate, for example, would collapse the meaningful distinctions *among* different places — along with the admittedly ambiguous but nevertheless real and meaningful distinctions generated by the binary of reality and its subset "fantasy." These distinctions among places are as central to the generation of human meaning and understanding as is the "inbetweenness" of the formally discrete yet interdepending variety of actors within any one place. This is explicit in Romanyshyn's (1989) already-noted argument that we count on things keeping their place to root our sanity. The collapse of differences among places also collapses the contexts and rooted practices that have duration and are also the enabling existential conditions for any human agency.[24]

Landscapes

Visual space is a frame for objects. Landscapes are distant and controlling; they instruct us in the ordering of space (Tuan 1977). This distance is related to vision. "Seeing does not involve our emotions deeply.... The person who just 'sees' is an onlooker... not otherwise involved with the scene. The world perceived through the eyes is more abstract than that known to us through the other senses" (Tuan 1974, 10).

If landscapes are the visible qualities of places, then the "imaginary places" represented by, for example, landscape painting or VEs are visually conceived and planned according to a logic that asks "what item" from "which category" will be placed within them. In contrast to places, landscapes (like the virtual military testing ground described in the introduction) are primarily *made* to be *seen*. They present the surficial quality of places even as they also reconfigure the environment according to a preconceived plan or program or set of notions that reflect specific understandings of the world, and the sociocultural role or position of the maker or makers of the landscape within this world. This is not to suggest that all landscapes result from unified instrumental action (Shields 1991, 24). Different overlapping histories and uses accrete over time at any one site. Landscapes inform the relationship between a people and a place and express communal values and interpersonal involvements (Relph 1976, 34). However, in their initial human construction, landscapes represent cultural practices writ large on the physical, nonhuman world.

Landscapes may be mundane or fantastical, sacred or profane. The material form of landscapes can range greatly. It might involve the manicured trees, flower beds, and lily ponds of an English country garden, and the bounded views a ha-ha establishes—from which an earlier English aristocracy might view a carefully tended and idealized English countryside as if looking at a painting. Or a landscape's form might be more metaphoric still and draw from an artist's palette, along with the acrylics, oils, and canvas with which she or he *visually* represents an environment that may be more or less natural or humanly constructed. This reworking of the material world into something picturelike helps strengthen the suggestion that landscape is the visible quality of place. The power of photography, and later visual technologies such as TV and video, also relies, in part, on a form of landscape production.

Landscape also operates to synthesize the political and the aesthetic. Using the landscape idea to understand VEs benefits from a distinction noted by Tuan (1978, 366), who accepts Philip Wheelwright's parsing of metaphor into two distinct constituents. *Epiphor* involves comparison. *Diaphor* creates meaning through juxtaposition and synthesis. Landscape is diaphoric, combining two different concepts, "domain" and "scenery." Domain implies political economy; scenery, the aesthetic. Landscape synthesizes the two. Synesthesia, Tuan continues, "is a probable condition for understanding and inventing metaphors." It "provides a foundation for the development of metaphorical thought" (367, 369). As representational landscapes, VEs operate within a similar synthesizing dynamic, but in a more technologized environment still than does the production of material landscapes. Distinctions between the technical dimensions and metaphoric powers of VR are difficult to sustain. The term "VR" captures this imprecision, and this is why I maintain the sometimes awkward parsing of VEs from virtual technologies. It is precisely at such moments when this distinction between technique and metaphor breaks down that "the effects of landscape are most vivid and the legacies of landscape most pertinent" (Harrison 1994, 212).

The planned vistas of Enlightenment-era English formal country estates, along with the paintings whose subject matter often depicts these planned displays of wealth and status, together constitute the landscapes described by Cosgrove (1984) and Stephen Daniels (1993). The scenic vistas, bucolic views, and pastoral nostalgia together create a representational space that is informed by absolute and relational conceptions of space. Like absolute space, these Enlightenment landscapes are planned, geometric, and meant to exert an effect. They are conceived as if the environment is a "blank slate," empty in advance of its being "peopled" by objects and things. However, these landscapes also display aspects of relational space. The overall energy created and excited by the planned interrelationality of the various forms within the frame is part of what gives landscapes their power. Further, these vistas and composed views subsequently excite human agency into a secondary representation of the landscape concept, executed in pictorial form. In both concrete and canvas forms, these representations spatially display an idealized set of meaningful social relations as an array. The frame that bounds a landscape environment—whether by Turner, Constable, Repton, or Brown—deflects attention from dynamic interrelating processes[25] and makes the space of representation over into a series of discrete events organized according to an understanding of space as empty until the performative modern subject confers or fills it with meaning—at which point relationality between the objects arrayed begins to exert influence.

For purposes of the present work, Western, European landscapes can be seen to act as a precursor to VEs. Their reliance on geometric perspective as a device directs privilege and focal power to the unitary viewer who can also be understood to depend on his or her eyes for a "point of view." In their reliance on the mechanics of vision, these landscapes help naturalize a distinction between the self and a living world that is rendered more abstract by inserting a distance between themselves and the viewer. Like eighteenth-century English landscapes, VEs are metaphoric worlds. However, unlike the fixity that early-modern landscapes both impose and suggest is a moral good, VEs, by virtue of their enabling software, have the potential to act on, as if by magic, the relationships depicted within their simulations. The frame encloses a landscape that at the scale of the individual viewer can seem ever changing and totally amenable to being reprogrammed. In the example offered in the introduction, as I walk through the landscape, it adjusts to accord with my virtual position within this representational space. As I fly down the street of the deserted town, the buildings on either side of me sequentially occupy a larger portion of the viewframe, then recede "behind" me and disappear from view. When I leave the town and enter the surrounding countryside, trees initially seen at a distance as small and indistinct grow larger and more identifiable as I navigate toward them. As well, at any point, "the enemy" may suddenly appear to challenge me, either as an attacking helicopter, soldier, or tank thundering down the road toward me from the "vanishing point." The program is capable of retaining information about sites I have visited, and may not, for example, "challenge" me with the same soldier in the same location more than once — given that it remembers my initial encounter with "him."

As human beings, we benefit from the interaction with contingent surprise that the nonhuman natural world offers. This is not present in a VE. Any contingent surprise "therein" involves a culturally inflected simulation. By means of their representational icons, users interact with one another. They may also do so with iconic representations of nonhuman things that have been designed and programmed according to human logic. All of this is of necessity limited in two ways. First, it is entirely human. Second, it reflects the cultural biases and sociopolitical assumptions of software writers, as well as the social relations within which they operate and the production of these technologies takes place.

W. J. T. Mitchell makes the arguable assertion that landscapes allow communication not only between people but "between the Human and the non-Human." Mitchell understands this as a mediation between culture and nature, continuing that landscapes are "not just a representation of a natural scene, but a *natural* representation of a natural scene, a trace or icon of nature *in* landscape itself" (1994, 15). Now, Mitchell is unclear here, collapsing several possibilities under the metaphor of "landscape." If he is writing only about, say, planting a row of trees with the intention of developing a pleasing vista, then he is on less shaky ground than if he is talking about a subsequent painting of these same trees, once they had attained maturity. In the first case, an element of the natural world remains. In the second, things have been entirely subsumed by their representations. Those early-modern landscapes resulting from physical manipulation of the natural world already prefigured what Leo Marx (1965) refers to in *The Machine in the Garden:* by virtue of their being an *icon of* nature *in* nature, an ideal conception of the natural is imposed atop, or conflated with, an existing natural setting in space. Virtual technology inverts both Mitchell's observation and Marx's succinct phrase. The garden — the taming, acculturation, and idealized representation of nature — is now in the machine, suggesting that technology has magically subsumed not only nature but the broader meaning of culture as well.

At least by the time John Ruskin was writing, landscape and nature had become almost interchangeable categories (Cosgrove 1984, 162). The perceptual difference experienced by the subject when in a virtual *surround* environment, as opposed to earlier technologies from landscape to TV, can be understood as the folding or enframing of place into the landscape concept. "All frames enclose (in the qualified sense of *surround*)" (Carroll 1988, 229). Landscapes depend on their frames, which constrain and give form to the meanings they are intended to convey. Landscape painting, television, and VEs all share in this understanding. What is different with VEs is the point of view of the subject. Before VEs, point of view was perceptually removed from the machine. With them, it is severed from its earlier spatial congruency with embodied location and extended forward into the frame to become part of the bounded object or cultural device. Users are repositioned as disembodied points of view — Hayles's "povs" — within the frame. In every direction they look, there is only landscape. The forward extensibility of the self necessary for this has been optically conditioned by the technology of perspective. This subject has become used to extending herself or himself conceptually "forward" along perspective's sight line to engage imaginatively with the view before her or him. The inside-outside distinction Cosgrove maintains existed for earlier landscapes is here relaxed, if not set aside. Furthermore, in a virtual space, at any one time, "subjectivity will no longer be uniquely linked to a particular perspective, for we can all literally occupy precisely the *same* perspective" (Simpson 1995, 158).

As a spatial model, or world-as-landscape, VEs metaphorically convert idea into matter, or imaginative sensibility into concept, while appearing to avoid dealing with any intervening sociohistorical material processes in achieving this "transcendence." Landscape becomes world, even as the concrete world of places continues to stand before us, outside the grasp of this metaphor.[26] World-as-landscape naturalizes the landscape idea, which Cosgrove (1984, 261) notes once acted as a bridge from a moral to a political economy. I find no reason why this is not the case with VEs, though the moral instruction might differ from what Cosgrove has in mind. Although earlier optical technologies demarcate between the viewer and their space of representation, they also contain the seeds of mechanisms (perspective and frame are two examples) that with lengthy technical refinement now blur the original demarcation and permit the (hybrid) cyborgian conjoinment of flesh and machine seemingly upon us.

Specific landscape forms represent specific cultural understandings constituted through specific social relations, times, and places. The older European landscape tradition may be formally exhausted, but the potential remains for its renewal in "other forms, other places" (Mitchell 1994, 14) — such as VEs. Today, landscapes are often treated as treasures, hang on gallery walls, or are accorded special environmental protection status. Like wilderness, they are perceived to be in short supply, and this suggests one reason for VEs' popularity even in advance of their widespread diffusion in commodity form. Their interactivity permits the making of landscapes; their polyvalency of form suggests "as many as you wish" at a time when the "real" thing — the natural world that is the partial referent behind earlier landscape production — for many seems to recede from view.

4. Sight and Space

The simulation of the deserted village under military siege described in the introduction—including such variables as the perspectives from which users view aspects of this environment, or how its landscape features are illuminated—constitutes a representational space that depends on an underlying foundation of number, language, and code.[1] No user's body can "enter" this digital space, originally premised on information exchanged as ones and zeros within a framework of Boolean logic. Visually oriented contemporary culture, however, which generally equates seeing with knowing, is open to suggestions being made by industry players and academics that VEs will actually offer a "space" into which postmodern subidentities of a once more unitary subjectivity might profitably relocate. Paul Smith (1988, xxxv) makes the spatialized argument that the individual subject occupies a series of *positions*—scholar, employee, partner, consumer, and so forth—into which he or she is momentarily called. The increased computational power that drives virtual technologies promises augmented possibilities simultaneously to act out these plural subidentities within iconographic spatial displays.

The Western experience of space is widely conceived to have a commonsensical association with seeing, the visible, and optics. How space is conceptualized has also been linked to metaphors of vision and physiological sight. Oral and written forms of language have been the vehicles used to assert these connections. In the move to text, the potential arose for a shift in the interplay between conceptions and experience of space that lessened the role of orality and increased that of visuality (Ong 1977). The more widespread dissemination of texts and development of a "book-acquiring public"—for Eric Havelock (1982, 57) the definition of literacy—augment the visual and spatial dimension of language. Yet oral and textual expressions, representations, and experiences of space retain a human cast. For Italian Renaissance humanism, man becomes the measure of all things, and visual and optical technologies such as the rediscovered Euclidean geometry, Ptolemaic cartography, and the camera obscura promote this emergent "worldview." Since this breathing of new life into classical metaphors of vision, how knowledge is understood to be produced and accessed has become increasingly imbued with visual metaphors, and informed by optical technologies to the degree that what cannot be seen may be argued, empirically, not to exist. This is not a simple causal progression. The invisible retains salience even if it is argued to be the purview of disorder. For mid-Victorians, "making things visible, making them emerge, became . . . a means of regaining control" (Beer 1996, 87). A similar dynamic underpins the Marxist idea of base and superstructure. The world of culture is a diverting, visible effect of the real power of economy that remains hidden and in need of excavation. So, too, with the Freudian notion of the unconscious — powerful drives remain hidden and must be brought to light. Extending Foucault, however, if power is strongest when veiled or hidden from view, then the trend toward visibilization at times reflects the political will of dominant and oppositional groupings to expose and undermine sources of power and resistance — whether the inmates of Bentham's panopticon or capital's power maintained through the diverting spectacle implicit in commodity fetishism.

In a similar manner, promotion of immersive VEs culls the history of optical technologies to fabricate a comforting metaphor equating the developmental stage of this new technology with that of, say, TV during the late 1920s.[2] Whereas TV does deliver a landscape to the seer, the heightened technical possibilities supporting VEs also allow an imaginary remaking and repositioning of this spatially individuated person's identity, by inserting his or her eyes *into* the view for the predilection of other users as well as himself or herself. In this chapter, I examine how the goal of *absolute or perfect clarity* — the recurrent philosopher's illusion (Dreyfus and Dreyfus 1964, xxi) — affects how space is conceived in VR. I try to maintain a distinction between (eye)sight and vision. We are endowed with a faculty of "vision" that has the meaning of a physiological mechanism, but also of hallucination, anticipation — as in prophetic vision — and metaphor. Sight refers to perception of objects by use of the eyes. Echoing Roger Bacon's late-medieval, post-Crusades wish for a clearer access to the divine, VR researchers believe their "world pictures" *model* a clearer access to reality. Whether a scientific procedure or a devotional practice, to model or envision environments as information means us-ing eyes and optics to see a world of variously copresent, discrete, con-tiguous, territorialized, and continuous spaces.

Space and the Eye

Seeing "has the effect of putting a distance between self and object. What we see is always 'out there' " (Tuan 1977, 146), and sight gives the self an exterior world (Tuan 1993, 96). Although one might argue either that sight is objective or that it has an extraordinary ability to stimulate fear, desire, or sympathy, nevertheless one sees as an onlooker at the edge of a mobile frame. Western conceptions of space, though not entirely reliant on the eye (an acoustical space[3] or the intimate environment of touch readily come to mind), have strong ties to vision. Sight requires light. "Light produces space, distance [and] orientation" (Blumenberg 1993, 31). Visual space is the farthest removed from our bodily sense and covers the largest "area" experienced by any sense (Tuan 1977, 399). Although to see is somehow to think and understand, "in particular, attendance to the purely visual region in the distance excludes awareness of the af-fective region [closer to the body]" (400). We gaze into a distant and open future. "What is ahead is what is not yet—and beckons" (400). Tuan de-tects this "forward" direction of vision—based, in part, on the preconcep-tual and bodily asymmetry of visual direction in a forward direction—in the "space" of progress. This "space" is also the conceptual destination sought by Roger Bacon to gain the synthesis of "illumination" and re-newed spiritual "direction."

Sight is unique in its ability to organize perception of space (Jonas 1982). Hans Jonas, defending the nobility and princely status of sight, argues that it is the only sense that establishes a "co-temporaneous mani-fold...which may be at rest" (136). Other senses rely on a temporal se-quence of sensations "which are in themselves time-bound and nonspatial" (136). Jonas notes light's unique ability to organize a contemporaneous spatial array. Real space, he asserts, is "a *principle* of co-temporaneous, discrete plurality" (138; emphasis added). Tuan (1977) connects space with freedom, and Jonas notes that the other senses "fall short" of sight with respect to the freedom they confer (1982, 139). Here Jonas is refer-ring to our ability to close our eyes, and he connects this physiological reality to free will, a link first made by Augustine, who noted our ability to refuse to see the light as well as our ability to open our eyes to dark-ness—implying on Augustine's part either an understanding of the difference between sight and vision or possibly a conflation of their meanings. Augustine is the first to conceive of "the road to God as passing through our own self-awareness" (Taylor 1994, 29), making an early link in the history of vision between the eye, self-awareness, and free will. In comparing the ear and the eye, McLuhan (1964, 144) theorizes that the "world of the ear" is "more embracing and inclusive than that of the eye can ever be." The cool, detached, and lidded eye, extended by literacy and a mechanical conception of time as linear progression, "leaves some gaps and some islands free from the unremitting acoustic pressure and reverberation."

Although the *organization* of human space is uniquely dependent on sight, space is *experienced* directly as having room to move, even as our spatializing faculties of sight and touch reveal it to us as being at a distance (Tuan 1977, 11). Vision's objectification of spatial reality is muted by our motility's relationship to space. Something of this relationship is suggested by the gestalt notion of the undifferentiated synthesis between ourselves and our lived world — a synthesis that VE designers seek to replicate. Sight offers us "a whole world or scene" (Rodaway 1994, 118). Paul Rodaway uses the same phrase as Tuan — "out there" — to situate the "geography of appearances" that sight establishes for the detached observer. He notes that although sight is often privileged as the most important sense, it is nonetheless limited. Appearances can deceive. Despite the identification of seeing with believing, vision is possibly the most easily fooled sense (124)[4] and of necessity relies on touch, one aspect of body motility. Space coexists with the sentient body, and original space is a contact with the world that precedes the thinking process. "Original space possesses structure and orientation by virtue of the presence of the human body" (Tuan 1977, 389). Jonas (1982, 141) writes that active body movement — what he reductively terms the "motor element" — "discloses spatial characteristics in the touch-object . . . the touch-qualities become arranged in a spatial scheme, they fall into the pattern of *surface* and become elements of *form*." The synthesis produced by eye-hand coordination is a spatial entity presenting "simultaneity through successiveness" (142). "In sight, selection by focusing proceeds noncommittally within the field which the total vision presents and in which all the elements are simultaneously available" (143). Although the word "noncommittally" is problematic — suggesting a minimal attention to the social relations that may have influenced in advance which objects the eye privileges within the spatial array it scans — this passage does suggest separate, related connections between sight and alienation, and alienation and freedom. These exist in tandem with the connection between vision and freedom theorized by Augustine and Jonas, and noted earlier. Because the simultaneity of image allows the seer to compare, interrelate, and detect proportion, objectivity is intimately connected to sight (Jonas 1982, 144).

Jonas believes in the "nobility" of sight even as he acknowledges our primary need to connect the sensations that flow from body motility (the hand's touch) with visual sense data gathered or received via the eye. He suggests that the resulting synthesis affords a higher-order sensory experience than either sight or touch on its own. I am partial to Jonas's phenomenological approach to sensation. He acknowledges the causal connections informing sight but also argues that it is the very nature of sight to eliminate these connections from its visual account (thereby, I suggest, contributing to its own exalted reputation). Because of this occlusion of its own causal genesis, Jonas argues that it is our responsibility to integrate the evidence sight provides *(theoria)* with "evidence of another kind" (1982, 147). This seems to be a reference to the fullness of a prediscursive corporeal intelligence. Nevertheless, his privileging of vision, and writing of the body as a "motor element," extends the Cartesian tradition of according primacy to sight in a way that conceptually privileges the eye over the human body of which it is still a part, and makes the eye a metaphor for the mind. Touch, the efficacy of which depends on body propinquity to what is physically contacted, comes to be seen as somehow more embodied and less perceptive than sight and is accorded a lower status of "helper" in a "hierarchy" of sensations that isolates the senses from one another and minimizes recognition of synesthesia and the interpolating role of metaphor in human understanding of sensory experiences (Tuan 1978). The eye-hand coordination that Jonas admits as leading to a higher-order understanding of the world is part of a corporeal intelligence based on an interplay of sensory differences that the Cartesian tradition pulls apart for analysis. This process makes it difficult even to acknowledge corporeal intelligence, for conceptually it has ceased to exist.

I argue for including the role of corporeal intelligence in understanding space. However, it serves no purpose to denigrate the eye, or to deny the brain's capacities devoted to processing visual information.[5] Yet I suspect that belief in the primacy of sight permits *theorizing* a state of affairs that does not exist: namely, a fully established binary between mind and body, and between an optical monarch and the remainder of the sensorium conceived to operate as a tributary system to an eye too readily conflated with mind.

Sight is the only sense in which advantage lies in *distance*—"the unfolding of space before the eye, under the magic of light, bears in itself the germ of infinity—as a perceptual aspect" (Jonas 1982, 151). Although touch and body motility reveal the potential of moving forward to the next point, and the next, and so forth, only sight continuously blends its present array into ever more distant "background-planes": there is a "co-represented readiness of the [visual] field to be penetrated, a positive pull which draws the glance on as the given content passes as it were of itself over into further contents" (151).[6] Given the primacy accorded to sight, and the connections it is possible to make between sight and the space of ever-opening distances it reveals to us, cultural naturalization of the assumption "space = distance" becomes easier to understand. If "seeing = believing" and "space = distance" flow from a "disinterested" sight's presumed preeminence in revealing and organizing space as a visual construct, then a related link between belief and truth, and objectivity and distance, can also be seen to have spatial underpinnings. Jonas concludes that distance gives us the idea of infinity. "Thus the mind has gone where vision pointed" (151). With VEs, embodied sight gazes upon machined vision. Eyesight is witness to the many visions or imaginings released yet contained by computational power.

Classical Greek Thought, Absolute Space, and VEs

"Space has a history" (Burgin 1989, 14). It is worth considering the line of inquiry, discussed in chapter 3, into how a modern understanding of absolute or empty space and its use within VEs might be informed by classical Greek theorems amalgamated into Euclidean geometry and its infinitely extensible 3-D space. F. M. Cornford's essay "The Invention of Space" (1936) is a philosophical history that asks how the Euclidean framework of geometric space was imposed on common sense. Euclid of Alexandria's (third century B.C.) *Elements of Geometry* codified several theorems developed independently of one another by Greek mathematicians during the preceding three centuries. Cornford's thesis is that belief in infinite space as a physical fact is traceable to Euclid's joining of the separate earlier theorems into a coherent whole, which required imposing an external grid within which all the theories could logically be situated. This grid coordinated and hence resolved incompatibilities made apparent when these theorems were brought together into a common geometry. Euclidean space "had no centre and no circumference . . . it was an immeasurable blank field, on which the mind could describe all the perfect figures of geometry, but which had no inherent shape of its own" (219). This framework is indebted to the classical atomists' belief in the existence of a boundless Void as a natural fact—the reasoned basis for endowing abstract space with physical existence.

Aristotle rejected the Void. His enduring authority, combined with religious opposition, suppressed the conceptual power of the Void until Galileo Galilei revived atomistic theory as a basis for his science and scandalized his pious contemporaries with an infinitely open space— the eventual consequences of which are to be seen in the victory of absolute space and spatiality over locality and places. In *Physics*, Aristotle comments on the spatial relationship between an earlier Pythagorean cosmology and the Void.

> The Pythagoreans too asserted that Void exists and that it enters the Heaven itself, which, as it were, breathes in from the *boundless* a sort of breath which is at the same time the Void. This keeps things apart, as if it constituted a sort of separation or distinction between things that are next to each other. This holds primarily in the case of numbers; for it is the Void that distinguishes their nature. (emphasis added)[7]

Heaven was a spherical universe, and the Void a name for the air that Heaven also breathed. This air, or Void, kept solid bodies apart and gave them room to move. Pythagoreans represented number by patterns of dots or pebbles arranged in shapes such as cubes and triangles.[8] Mathematical thinking was geometrically represented. "The Void which distinguishes their nature is the blank intervals between these units, or the gaps separating the terms in a series of natural integers" (Cornford 1936, 224). Pythagoreans held that physical bodies actually were numbers. Visible, tangible bodies are aggregated from a plurality of units equally held to be the points of geometry, the atoms of bodies, and the units of arithmetic. Both separation and multiplicity of these units are maintained by the Void that is always everywhere between the surfaces of different bodies (225). In other words, though culture (which produces number theory) at least defines nature (a body is really a number), a conceptually messy Void—positioned between the natural air both humans and Heaven breathe and the cultural "blank interval" that maintains numerical purity—is required to maintain this definition. A number-governed universe can be deduced by pure reasoning. Observation is of little use (Thuan 1995, 11).

Religious belief permeates these understandings. The planets and celestial objects were not expected to move in erratic, undignified ways.
Geometric explanations (theoretical reasoning) were sought in order to
"save the appearances" of the (empirically observable) planets, which
seemed to move in erratic fashion when in a retrograde phase of movement relative to the Earth's motion. A circular, uniform motion was the
"correct" way of the gods. "For the Greeks, geometry was the morality of
motion" (Walter 1988, 186). One of Plato's sayings was "God is always
doing Geometry" (Kitto 1964, 193).

Cornford argues there was no intention on the part of early Greek
philosophy to conceptualize something so scandalous as an infinite extensibility beyond Heaven. Instead, the theorization of infinite spatial extensibility hinges on the meanings, implications, and *translations* of the
word "boundless" found in the later writings of neo-Pythagoreans and
Atomists such as the Roman Lucretius (99–55 B.C.). Pythagorean theory
drew from the thought of Anaximander (610–547 B.C.) and his pupil
Anaximenes (588–524 B.C.), who taught that the cosmos was composed
of materials[9] taken from a "boundless" body encompassing the world.
This body was the air Heaven breathed. Boundless, however, does not
mean shapeless. For these ancients, the circle and the sphere were "boundless," as is a simple, unadorned finger ring or the edge of a coin, for neither has a discernible beginning or end. For the dramatist Euripides
(480–406? B.C.), the ether is boundless because it is round and embraces the earth in its arms (Walter 1988, 227). Cornford asserts that the
"boundless" Void of early Pythagorean thought implied spherical shape,
that the concept of boundlessness joined a spherical sense of the finite
with the unbounded. This accords with what Victor Burgin (1989, 15)
calls the space of "common sense": "The horizon appears to encircle us,
and the heavens appear, to the eye, as vaulted above us." Trihn Xuan
Thuan (1995, 11) notes the geometrical harmony of Pythagoras's *kosmos,* which was based on the most "perfect" mathematical form — the
sphere. Parmenides later denies a boundless space extending into infinity beyond Heaven because no reason existed for imagining such an unoccupied spatial infinity. Instead, space had a spherical form with a center and circumference.

To free the mathematical Void from physics required distinguishing it
from air (Thuan 1995, 229). Parmenides' refuting of the Void was taken
up by the Atomists Leucippus and Democritus (460–370 B.C.). They admitted the Void was nothing, or "not-Being." Reality alone existed in
atoms, which were "that which cannot be cut finer" (Kitto 1964, 200).

However, they set the nothingness of the Void, believed essential to allow for movement of objects, within a binary and opposed it with *something*, with matter or body. Detached from its meaning as air, the Void of immaterial, geometrical space was now a single continuous medium. The Void is freed from its origins in nature and, in being made over into something entirely conceptual, moved to the realm of culture. It is also freed from visual empirical observation and moved to the sphere of pure reason. By Simplicius's account, "there is a void, not only inside the cosmos, but also outside—a thing which clearly will not be 'place,' but something with an independent existence."[10] This reads similarly to Latour's understanding of the independent factual reality of the vacuum brought to light by Boyle's intervention. Henceforth space has neither circumference nor center but rather an unlimited extent more in keeping with the scientific needs of geometers.

When Lucretius translated the meaning of the Void into Latin, the connection between boundlessness and the O, or perimeter, of a circle was minimized in favor of asserting a link between boundlessness and "#," or gridded extensibility, as this second icon implies. There is an interesting connection between neo-Pythagorean and Neoplatonic belief. The neo-Pythagorean identification of divine reality with the One, and all other realities as emanating from the One, is most likely the basis by which Neoplatonism makes the same assertion (Reese 1980, 471). For neo-Pythagoreans, numbers preexist in God's mind, and number is the essence of reality. As such, number is the essence of a preordained geometric space. For Neoplatonists informed by neo-Pythagorean thought, the ray of light, as an *emissary* from an originary God, is in essence number—a conception of reality replicated in VEs, "wherein" light and the simulations it arrays are controlled by number. At base, then, an array of light constituted in divine number is a communication from the nonhuman to the human sphere. Number as part of culture is really God given, hence not only its distinction from culture but the implicit permission to inscribe it over natural bodies and things. The association of number with transcendence is strengthened by connecting number to light and the power it bestows on Earth.

This partial account of what might be called the "element complexes" that inform the genesis of Greek absolute space suggests that specific traditions and beliefs continue on in new virtual technologies, which in themselves are sometimes argued to be in opposition to narrative or history (Bukatman 1993). For virtual technology to achieve cultural salience, its designers and builders need to assert that it constitutes a distinct, eminently real, empirical reality, not only in its being modeled as an absolute space, but more in the claims made regarding what this well-lit geometric space will provide and even effect. At the same time, VEs instantiate a relation between events and things and do so in a very individualizing fashion. Everything is depicted in a relational manner — people, things, and backgrounds included. All such relationships are subject to modification via either language or the user's wish. In a VE, spatial planes can seem to merge with objects, and the reverse also appears possible. Displayed objects exist in a two-dimensional interrelationship with other objects. Each has something of these other objects within itself by virtue of its relationship to them. NASA scientist Stephen Ellis can note the Cartesian framework or absolute quality of a VE spatial display on the one hand yet also write that "pictorial communication is seen to have two directions: (1) from the picture to the viewer and (2) from the viewer to the picture." Further, "a picture is produced through establishment of a relation between one space and another so that some spatial properties of the first are preserved in the second, which is its image" (1991b, 22). Ellis seems to suggest that a relational notion of space is important at the scale of individual experience within a VE. This contrasts with the absolute space of the virtual grid on which this experience is displayed and suggests something of a retreat from the modern pictorial tradition, and a return to a medieval one — one in which symbolic importance was not yet so subordinated to spatial regularization.

Virtual technology scientists routinely refer to the "space" within a VE as Cartesian (see Ellis 1991a; Kaiser 1991). The "environmental space" or "field of action" within a VE may be defined "as the Cartesian product of all the elements of the position vector over their possible range" (Ellis 1991a, 4). Although Ellis theorizes the self as a distinct "element" *within* this space (as it provides a point of view around which the environment is constructed), this self is a kind of relational "first among equals." "The balls on a billiards table may be considered the content of the 'billiards table' environment and the cue ball combined with the pool player may be considered the 'self' " (4).[11] For Ellis, the representational space of the "billiards table" environment is composed of the things within it in a way that synthesizes Newtonian absolute and relative space, a Cartesian notion of extension in a world of matter, and a relational concept of space as determined by the system of distance relationships between things in space (or environment). Ellis's assertion that pictures depend on establishing spatial relationships so that spatial properties from one site might transfer as visual images to another suggests the essence of absolute space — where the nature of objects is irrelevant. However, the abstract "Cartesian space" he describes — like relative space — is grasped as a system of relations between one or more sets of objects.

Yet as Held and Durlach (1991) note, aspects of users' subjectivity are both in and on the display, that is to say, both *in* and *on* this representational space. Stated otherwise, this space both subtends subjectivity and forms a part or place for it. In a VE, two dynamics are at play. The subject agrees to move conceptually into the virtual world or spatial display. In return, the technology provides him or her with a point of view that has the mixed existential and metaphysical status of a location for, *and* essence of, subjectivity. Vision and sight seem fully relocated to this abstract site/subjectivity. Two related but contradictory results follow. The relational fluidity of identity and space in a VE affirms the suzerainty of the modern subject constituted according to vision and possessed of an inner light. Yet this is possible only by having situated fluidity within an abstract matrix predicated on an underlying absolute grid, made to recede experientially because of (1) the collapse of space between viewer and machine, and (2) the ongoing high cultural capital accorded to fluidity and its promise of polyvalent pleasures and material rewards.[12]

The Visibility/Invisibility of Space

The underlying invisibility of information technologies and telematics may render them less accessible to moderns whose identity and sensitivity are oriented to the visible. One of the dynamics driving the dissemination of virtual technologies has been to provide an interface of visibility to allow perceptual access to the invisible dataspace of information flow.[13] Digital space has been theorized as a parallel world "with many more windows into it than the humble T.V.'s familiar vacuum-sealed, plexiglass porthole — and yet which is strangely *invisible*" (Wark 1993, 142). What is the purpose of 1980s stealth technology, for example, other than a military exercise to regain or assert the upper hand or competitive advantage by the use of invisibility? U.S. stealth flight technology is designed to overcome the widely shared perception that power is most vulnerable at its moment of application because it must take on visible forms. The F-22 stealth aircraft partially circumvents this vulnerability by being difficult to see or detect on radar. We are constantly urged in myriad ways to demure to the ideology of "seeing is believing" that is increasingly coupled to semi-explicit assertions that representations are virtually coeval with that of their referents. In such contexts, the existence of stealth as a technology — even though the plane is visible to human sight—is capable of being questioned, a destabilizing eventuality believed by stealth's early proponents to advance their interests. Given current epistemologies, stealth puts into question power's very existence (Virilio 1989) and suggests something of the rapidly changing relationship between space and visibility that digital/optical technologies of visibility/invisibility are thrusting upon the lived world. This ironic questioning can take place at the same time as VR technology renders a vision available to sight that was previously invisible.

Cyberspace, however, never was invisible at all, in the same way that a painting is not invisible before rendered by the artist. It is important to remember the power inherent in the collapse of the frame. With respect to immersive VR, issues of invisibility and nonexistence get conflated in claims that VEs manifest something that otherwise could not have been seen. In a VE, something is seen, but the image could not exist except as an idea or vision to the "mind's eye" without humans engaging the technology's materiality. In a sense, if the argument that VEs visibilize cyberspace forms part of a broader support for "seeing is believing," then the argument is self-defeating. Implicitly, the issue of fear around what cannot be seen gets raised, and the power of invisibility is made stronger through assertions that these technologies manifest what could previously not be seen. Some*thing* nonexistent has seemingly been made visible. For those seeking certainty through optics, however, the power of invisibility—and all that remains unavailable to vision (who knows how large a field that might be?)—remains undiminished, and the technology may ironically confirm already held anxieties about the invisible realm.[14] If Bentham's fear of ghosts and invisible specters informed the panopticon (see note 16, this chapter), did the device, or his intention to live within a glass house full of mirrors and optical devices, really exorcise his demons? Or did they, instead, serve as ongoing anodynes to a fear that forever haunted him because, in part, it was cemented in a foundation of empiricism? To the degree that "seeing is believing" becomes accepted ideology (and this is never fully the case), the subject who senses the invisible but cannot fully acknowledge that it is really there may experience a crisis of belief. Or rather, identity may be threatened, and recourse thereby made to phantasmic uses of technology such as evinced in magic lanterns, the cinema, and VR.

A Shadow Theater of the Invisible

Bukatman argues that telematics "are invisible, circulating outside of the human experiences of space and time" (1993, 2). In theorizing a space *apart* from human experience, and by extension a technology free of human constraint—a separate language of "outer" space—he unwittingly illustrates the continuing cultural saliency of the age-old belief in an "absolute space" separate from embodied reality. The issue of invisibility, therefore, is also worth noting because of the connection between space and vision. Cyberspace, a phenomenon partially reliant on absolute space, is invisible, or at least initially unavailable to visual perception. This has demanded VEs, as a subsequent overlay of technology, to make experientially accessible this "space" of #, language, and metaphor.

Electronic data flows are invisible to individuals. Without an interface device, one cannot "see" into cyberspatial parallel "worlds," yet the latter are new centers of power. NASA scientist Mary Kaiser (1991, 45) writes of the "unique visualization tools" that will allow users to adapt the "scales" of time and space by offering vantage points "not actually achievable to observers, and by making invisible forces visible." However, Kaiser continues, visibilizing the previously invisible will not only permit exploration of new worlds but also "embody spatial metaphors which exploit our propensity for spatial thought to understand structures from nonspatial domains better" (46). For Kaiser, spatial thought is visual thought (43). Every phenomenon that has previously eluded sight's purview will soon be made apparent to the eye courtesy of spatial displays.

VEs are the spatializing or visibilizing of the digital language used in virtual technologies. People cannot see into, much less physically enter, the conduits and data flows that are now the ironically rhizomelike centers of power for economic and military institutions making use of this space of flows. Hence the rush to visibilize data flows as VEs, the rush to the money. For all the hype about the entertainment and educational potential of these technologies, they are the new sine qua non, transnational site for in-house communications between the military, financial, and industrial arms of dominant cultural forces. If invisible data flows have become a nexus of power, then many people will be drawn to find the means to experience these flows of information-as-knowledge in a sensual fashion by conceptually merging with on-line VEs and other as-of-yet unimagined optical procedures. This eventuality will demand a further objectification of individual bodies. The "center" or "space" of individuality represented as a point of view moves forward to extend its presence not only "across" a gulf separating subjects and objects but "into" this gulf or "space" of distance, of which this point of view—like the "pov" Hayles identifies as one of *Neuromancer*'s cultural innovations—wishes to be a part. Although the form of this dynamic may seem novel, issues of invisibility notwithstanding, politically there is little new here. In a world organized according to the apparently horizontal and democratizing logic of electronic communications, this is what it now takes to get close to a contemporary "center" of power.

Bridging the Space of Distance: Two Models of Subjective Vision

At least since Filippo Brunelleschi, Leon Battista Alberti's *de Pictura* (1436), and the wider Italian Renaissance rediscovery of perspective, a key aspect of the subject's power of sight has been possible to theorize as extending along a sight line running between subject and object. This sight line spatializes human relationships by privileging the distance it establishes between the luminary eye of the unified subject and the object of its view. From this modern subject's point of view, objects of desire, like the future, are always ahead of him or her. No matter how close an object may seem, it remains at a certain tantalizing remove unless the subject takes physical action to make direct contact with it. Yet the subject's conceptual integrity depends on the continued existence of its discrete, spatially removed other, even though this subject is equally objectifiable by other humans also acting as objectifying individual subjects. The metaphoric and concrete distances established and maintained by a sight line between subject and object support modern alienation, along with a related and more positively connoted ability for criticism and objectivity. Jonas argues that the manner in which sight spatially arrays copresent things into an indefinite distance suggests that if there is any "direction" to how objects are spatially organized, it is "away from the subject rather than toward it" (1982, 136). The physiology of sight itself suggests the unattainability of the object. Furthermore, "it would not be correct to say that in sight the distant is brought near. Rather it is left in its distance, and if this is great enough it can put the observed object outside the sphere of possible intercourse" (151). The sense of moving forward contained in the notion of progress, itself a visually dependent spatial overlay onto time, together with sight's particular construction of distance, prods the subject to link up with the object of vision, with the "other," which must nevertheless continue to maintain its (visual) distance from the subject. The separate identity of the object also depends on its dialectical position at the other end of a sight line that must never disappear or stop being maintained. Although the possibility of surrender to the other is minimized and desire left unfulfilled, the existence of a body to which the eye connects or harkens back, and the materiality of the object that it seeks, are not overtly disavowed by this schema. This is because the asymmetry of our bodies has been difficult to forget or explain away. An individual's vision extends outward unidirectionally. We may change our points of view, but only by shifting our bodies, the most obvious yet subtle example being moving the head from side to side. The subject-object dichotomy maintained by the modern sight line also mirrors the modern Western dualism of the body-mind relationship.

In addition, this sight line traces the space of distance that communications geographers, for example, have pronounced "collapsed." In my earlier discussion of Carey's (1975) understanding of communication, I suggested that a Western belief in "space = distance" has led to the conflation of communication with transmission, for space is always *seen* as an impediment to be crossed and rarely as a field on which subjects might gather together. In the sense of visibilizing or spatializing the concept of a "progressive future," conceiving of space as distance implies that we may never get to where we are going, but we may glimpse an imaginary destination before us. As subjects we may not conjoin with the object in view if we are each to maintain our visually dependent, distance-separated, unified identity formations, but we are not constrained from communicating across this space with one another. I am not proposing that we only see in the focused unitary direction suggested by the following depiction of a line of sight, where x is the subject and y is the object in view:

$$x \longrightarrow \text{line of sight} \longrightarrow y$$

Rather, the sight line between subject and object has a vertical span of 150 degrees and a horizontal one of 180 degrees—the actual range of human vision. We are capable of a wide range of vision, much of it classified as "peripheral." However, I think it likely that the way in which modern Western individuals deploy sight has been culturally influenced by the one-point perspective model that the simple diagram also approximates, for such a perspective is uncluttered and highly focused. What exists to the "left" or "right," "above" or "below" this reductionist, atemporal depiction of the human line of sight—what remains unseen because it is beyond visual perception—has nevertheless been subject to human conceptualization in visual terms. From "the mind's eye" it is possible to envision what one's point of view might see if it were in other places—to have an image of a relationship between vision and sight from a metaphoric "interiorized stance" not unlike that ancient heavenly place "on high" from whence descended classical Absolute light. This spatial understanding informs visual techniques and technologies such as cartography, landscape-as-idea, the dissemination of printed texts, the production of perspective painting, and aspects of the history of the modern novel up to and including science fiction as a speculative utopia.

In *The Civilizing Process*, Norbert Elias (1968) identifies the construction of a fictional and invisible "wall" — a spatial metaphor underpinning modern identity that demarcates the interiorized self from objects and other individuals outside "it." Over the past seven hundred years, this "wall" has helped insert a pause between the brain's command and the hand's carrying out of this command, a break inserted into the seamlessness of eye-hand coordination. This separation permits the rise and ongoing refinement of modern social relationships and capitalist economies based on differentiation of labor skills. In an economy based on division of labor, it is unproductive for individuals dependent on one another for goods and services to kill each other spontaneously during heightened emotional states. This "wall" helps create a critical distance and a cooling off period that minimizes bodily harm and social disruptions. Elias writes of

> an eternal condition of spatial separation between a mental apparatus apparently locked "inside" man . . . and the objects "outside" and divided from it by an *invisible wall.* . . . The act of conceptual distancing from the objects of thought that any more emotionally controlled reflection involves . . . appears to self-perception . . . *as a distance actually existing* between the thinking subject and the objects of his thought. (1968, 256–57; emphasis added)

VEs are an attempt to supersede the modern constraints imposed by this distance or "wall" between subject and object, and, by extension, between subject and society, which comes to be experienced as "external" to, or other than, the subject. Users, immersed and interactive, can forget momentarily that they are interacting with representations of other people and things, and that the transparent screen in many ways reifies Elias's wall even as it appears to offer a way to vault over it. VEs may be thought of as bringing vision to sight: they seem to offer directly to physical sight the multiple, scattered visions resident within "the mind's eye." Figure 6 models aspects of identity in virtual space and how this differs from the older modern sight line sketched earlier. The subject-self *(A2)* wears an HMD that wraps the viewing screen in close proximity to the eyes. On the virtual side of this screen are the subject's extended points of view *(B1–B4)*, illustrated as disembodied eyes with sight lines extending outward from each. Also shown are the data, commodities, gaming, virtual teleconferencing, and so forth, that constitute the utilitarian appeal or "functionality" for users in VEs. Researchers are also working to develop virtual technology that would allow the eye to take multiple simultaneous positions. Figure 6 diagrams a possible array of spatial relationships within such a VE.[15]

In this model of cyberspace telepresence, the subject's eye remains linked but detached from his or her body. This state of affairs may be imagined as facilitated by the eye's extension along what I will metaphorically term a coaxial cable that may be imagined as a flexible sight line operating within a full six degrees of freedom. As in an out-of-body experience, this cable connects the subject-self who remains grounded in subjective and embodied reality across or behind the interface/frontier/dialectic of the screen with the emergent imperium of cyberspace. This "coaxialized" extension of self-identity into virtuality might seem to offer the illusion of an emotionally satisfying alternative to the inability of the visually conceived unitary subject to join with its dialectical "other."

Virtual points of view *B3* and *B4* suggest how it might also be possible for separate images, or avatars, of the self to face each other as seemingly discrete entities, and also for that part of the self remaining on this side of the interface to watch both, as in "I see myself seeing myself." This relationship may also occur in "real" space and is shown by *A1* gazing at *A2*. Virtual self *B2* suggests the potential for the disembodied point of view to turn back and gaze upon its body (*A2*), which may appear to it as an "other" or as a shell. For the self, whose spatial coordinates now are split, these relationships beg the question of "where" its identity is located. Identity begins to exist and situate within a schizophrenic dialectic operating, as it were, unto itself, ironic or incoherent as this might first appear.

I wish to make two related points that suggest that the alienation users might seek to escape or overcome within VEs ironically may proliferate. First, a close reading of Elias's description of "an invisible wall . . . as a distance actually existing" suggests that not only is there a subject or "mental apparatus" locked "inside" the modern individual along with the object resident on the other side of "the wall"; there is also a second aspect of subjectivity that has stepped back from what it views in order to be able to identify the self-conscious subject-(wall)-object relationship proposed by Elias and constructed by VEs. This is suggested in the viewing relationship between *A1* and *A2*. This second aspect may be thought of as you, the reader, who looks at the diagram and imagines yourself at various positions within it. This duality between interiorized aspects of

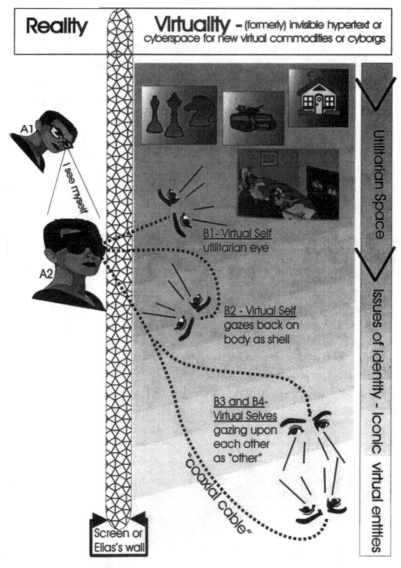

Figure 6. Aspects of identity in virtual space. Illustration by Liz McKenzie.

a supposedly unitary subjectivity is sometimes understood in terms of "I" and "Me," or as present to oneself. This modern sense of "consciousness" long precedes the development of virtual worlds. However, the fracture of self-identity implicit in the relationship "I see myself" is seemingly multiplied in "I see *Me* seeing myself seeing myself"; and this fracture gains support within a cultural context expressed by the phrase "I like to watch." Second, figure 6 therefore also models a variation of the Hobbesian Author-Actor binary discussed in the introduction and chapter 1. The parts of the self that represent the Actor—all the *B*s, or virtual selves, but also *A2* in relation to *A1*—venture forth as iconic, or masked, players, while the Author *(A2)* with respect to virtual selves *B1–B4*, but also *A1* in relation to *A2*, monitors and possibly guides these more public aspects of the performing self. At the same time, *A2*, for example, may imagine itself conjoined with the *B*s in a technologically mediated telepresence. The Hobbesian Author or "I" remains detached, however, from the Actors it controls "at a distance," who in turn control the representations of self that the entirety of what is implied by "present to oneself" also seeks to join.

For many users, however, the iconographic communications of a VE may not feel representational or mediated. VEs flooded by light offer more experientially real simulations of sight's physiology than was possible with earlier visual communications technologies. With sight, "the percipient remains entirely free from causal involvement in the things to be perceived" (Jonas 1982, 148). This implies that image is *being* in and of itself before representation. The power of image is partly to suggest that regardless of how culturally constructed a particular image may be, somehow it precedes human intentionality. Extending Jonas, image, like Kantian space, enjoys an a priori status. To be sure, this power of images exists in the real as well as the virtual world, but the latter's completely light-dependent immateriality enhances the particular nature and claim of this power. Image "becomes essence separable from existence and therewith theory possible" (149).

To believe that one might achieve transcendence from the bonds of subjectivity and join with the "other" in representational space would be to situate oneself conceptually in a space that owes a debt to Jeremy Bentham's panopticon (1787). Conceived in the same year as the panorama, with which it shares many conceptual and spatial similarities, the panopticon, or "inspection house," is a succinct modern metaphor for how visually dependent spatial arrangements between subjects and objects are conceived and put into practice. The panopticon models territorial segmentation and spatial strategies of control along visual lines. Anticipating multipurpose buildings peopled with plural identities, the device— "a prison governed by the eye" (Oettermann 1997, 40)—was intended to "punish the incorrigible, guard the insane, reform the vicious, confine the suspected, employ the idle, maintain the helpless, cure the sick, instruct the willing" (Sack 1986, 183). Its inhabitants were variously "a madman, a patient, a condemned man, a worker or a schoolboy" (Foucault 1979, 200). Like the lidded, blinking eye, the panopticon's spatial strategy of containment and surveillance can be turned on and off as required. Its multiuse program links spatial segmentation to the segmentation of identities already well under way by the late eighteenth century.

Panopticized subjects are constantly exposed to an exterior gaze, and if the individual is to normalize this situation, this gaze must be internalized. Any secret that might be seen is eliminated, and the panopticon represents an obsession with what cannot be perceived (Hollier 1984, 14).[16] Visibility thus becomes confinement, promoting isolation, inwardness, and the self-discipline reflected in an acceptance of Elias's "wall" as a "natural fact." The panopticon is the origin of the perceived inner division of a consciousness continuously engaged in self-monitoring (Foucault 1979) — the being "present to oneself" discussed earlier (and outlined schematically in the relationship between *A1* and *A2* in figure 6), and resonant with the gestating "light" of inner subjectivity discussed in depth in the following chapter. In the pixelated light world of a VE, where "all the world's a display," visibility is paramount. An objection may be raised to linking what Foucault refers to as *panopticism* (or what Lieven de Cauter [1993, 21], critiquing Foucault's inadequate consideration of links between spectacle and discipline — and hence the panorama to the panopticon — terms the panoramic eye) and virtual technologies on the grounds that both subjects and objects are invisible in information space and that virtual entities are multiple and dispersed rather than bipolar (see Schulz 1993, 438). Yet the omnipresence in virtual technologies reflected in the ability for data to be transcribed and reviewed by whomever has access and power to do so is very similar to the panopticon. Too close a focus on the bipolar nature of the guard-teacher and the inmates-pupils minimizes consideration that the guard-teacher is not paramount. "The Prison director's office or lodge was placed inside the guard tower in such a way that he, too remained out of sight. Through a series of slits in the lodge walls he would be able to watch not only the prisoners but also the guards watching them: a hierarchy of total visual control" (Oettermann 1997, 40). We may be sure that the director reported to higher authorities. Further, whereas the information space of ITs is invisible, it should be recalled that two hundred years of internalization of what it means to be randomly watched by authorities is brought to the table with respect to how ITs are deployed. The contemporary workplace is rife with stories of employers reading employees' electronic correspondence at random and unannounced intervals. This is the discipline of the panoramic gaze applied to the workplace in the name of efficiency.

Virtual "Pleasure"?

Although VR *promises* a potentially infinite number of points or different recombinations of depictions of environments dispersed throughout the field it models — as opposed to the panopticon's unitary and centralized point from which surveillance of subjects is conducted by "middle-level management" — both VR and the panopticon isolate users physically and leave them uncertain or unaware of observation, transcription, and review by others. VR does not offer a range of individual or collective presences with which to interact; it offers a range of devices and computational patterns. Physically isolated, VR users consent to monitoring by tracking devices at the same time as they review the performances of their own representational avatars or "puppets" on the virtual display-stage. In a sense, the operation of power within VR becomes somewhat automatic — users become both perpetrators of, and subject to, the power reflected by and circulating within the virtual environment. More so than with text-based Internet technologies, or broadcast media, and the sometimes ambivalent and segmented practices of viewing and listening encouraged by TV and radio, VR also helps produce the subject's consent and self-disciplining because it requires users to wear the technology on their bodies. As such, it trades on the current trend toward wearable technologies such as beepers and cell phones and portable E-mail devices. An employee wearing a paging device is always available. The "optical regime" also retains power through spatialized control mechanisms that were always as important as visibility and the functionality of the panoramic gaze.

In relating VR to panoptic devices, I am emphasizing here the surveillance possibilities of optical technologies (see Lyon 1994). Much contemporary writing on VR, however, emphasizes the polyvalency and pleasure — the "free play" — that is the potential, the premise, and indeed the hype that fuels interest in the technology. I suggest that pleasure and surveillance exist within a dialectical relationship within VR that need not be theorized as oppositional. A narrow emphasis on articulating VR to pleasure alone only proceeds if one has agreed to the ethics of the technology that demand users agree to comport themselves within the machine's dynamics, a central component of which remains the ability to review and transcribe users' activities and choices. An emphasis on the pleasure of VR-as-text has the potential to deflect attention from the surveillance capacities of the technology. Hayles (1993b) argues that the transformations "that take place as the body is translated from a material substrate into pure information" shift cultural emphases from a focus on physical presence to one more concerned with computation and the patterning that it supports and demands. "Working with a VR simulation, the user learns to move her hand in stylized gestures that the computer can accommodate. In the process, changes take place in the neural configuration of the user's brain, some of which can be long-lasting. The computer molds the human even as the human builds the computer" (90).[17] With respect to pleasure and surveillance, Hayles's assessment of VR reads well against Oettermann's linkage of the panorama and panopticon as mutually reinforcing capitalist technologies. "As 'schools of vision,' the panorama and panopticon are at the same time identical and antithetical: in the panorama the observer is schooled in a way of seeing that is taught to prisoners in the panopticon" (1997, 41).

Depending on one's political assessment of the relationship between production and consumption, the dialectical relationship between pleasure and surveillance instantiated by VR can be positioned, for example, as an evil triumph of surveillance, a progressive victory for personal choice, or an amalgam of these polar opposites. In chapter 6, I make use of Roger Caillois's (1984) theorization of psychasthenia. Caillois's understandings in this regard have been the subject of considerable academic interest (Olalquiaga 1992; Grosz 1995; Hillis 1998). Rather less attention has been paid to his writing linking production and consumption to the issue of mastery and slavery. Caillois argues what is tantamount to a homology between consumption and slavery, and he would deny any human attribute of pleasure on the part of those who control the modes of production. It is, instead, what can be gleaned from how Caillois positions the perception of the producer as he or she witnesses the consumer in action that is germane to the present discussion of (optical) surveillance and pleasure. Caillois's theory of consumption/slavery and production/mastery posits that "the slaves imagine that pleasure is 'the highest goal of freedom.'. . . the masters know that, quite on the contrary, it is 'the main gate to slavery'" (Hollier 1984, 7).[18] Caillois's position is controversial, and if one approaches an analysis of VR from the vantage point of *either* production or consumption, the dialectic I am proposing the technology suggests between pleasure and surveillance is unavailable, for one will see the technology either as entirely pernicious (akin to a "reverse" technological determinism) or as a utilitarian "conduit" toward happiness as the highest goal.

In keeping with my goal to avoid the sterility of determinisms, I introduce Caillois's observations because he points to both pleasure and surveillance in his assessment, even though he does not make links, for example, between the producer/master who may take pleasure from his or her surveillance of the "slavish consumption" by others of the representations and products he or she produces, perhaps with the help of their very labor. I am reminded of Norman O. Brown's description in *Love's Body* (1966) of individuals as they approach the object of their desire. He uses the metaphor of the rose. As the seeker approaches the rose, he or she first exclaims, "Roses, roses." Subsequently, the seeker, excited by what has first been seen, grasps the bloom by its stem, only to exclaim, "thorns, thorns," as he or she comes to realize that work and struggle on the part of those who might be termed truly active readers also attend any worthwhile process of "coming to know" that may also have pleasurable effects. I am not suggesting that thorns equal surveillance, but I am arguing that too often the pleasurable effects deflect attention from, for example, the scopic mechanisms of power that inveigle VR as a technology and social practice.

Continual surveillance of body motility is one price paid for gaining access to an "influencing machine" that renders users as commodity forms — icons, puppets, and avatars — within which they feel less responsible for their actions (the pleasure of becoming a commodity), yet also exalted to act as if a God-machine made of light. The ghost of the panopticon haunts VE theorist Jay Lemke's (1993) description of the relationship between sight and access in virtual worlds. "Material-world data will include 3-dimensional recordings of human activities and events that we can enter, move around in to see from any point-of-view, touch, and manipulate in every conceivable way, as data."[19] Yet VR offers a restricted range of experiences in that it is less about a range organized according to "as many as there are people," and more according to "as many as there are different VR devices."[20]

In the real world, if I want to touch something physically, I must reach with my body to do so. In a VE, sight is partially detached from this motility (or substance requirement), and what is experienced as a kind of "flying sight" or adjustable bird's-eye view can seem momentarily to breech the distance established by modern subjectivity. Yet it hardly needs stating that VEs cannot collapse physical distances between sub-ject and object or things more generally arrayed in space. If anything, they may reify psychic "distances" in the real world, with users spending in-creasing amounts of time in all forms of cyberspace to the detriment of maintaining their embodied social relations. VEs, as part of a continual juggernaut of progressive technological refinement and change,[21] sug-gest that a further shrinkage of the subject-object dichotomy may well be "on view"... just past the next blind spot... almost on the horizon...

Sight and Motility

Exploring the physical space we see means moving our bodies through this space. Visual displacement does not offer the same experience of mo-tion as does movement through space. The dynamic character of active motility discloses geometric relationships between things in space (Jonas 1982, 155). Scientific study of optical perception has arrived at similar conclusions. Visual stimulation independent of self-produced movement leads to abnormal sensory-motor coordination. Richard Held (1972) con-ducted a number of experiments on animals and humans in which sub-jects were moved about in space by mechanical devices. Newborn kittens passively transported in a gondola round and round within a spherical room received the same visual stimulation as active kittens in a control group. "Active kittens developed normal sensory-motor coordination; pas-sive kittens failed to do so until after being freed for several days" (373). Similar tests on human subjects revealed links between visual and mo-tor processes in the central nervous system. All these experiments suggest to Held that "close correlation" between signals from the nervous sys-tem leading to body motility and consequent sensory feedback causally related to movement is crucial in spatial adaptation to environments. "The importance of... self-produced movement derives from the fact that only an organism that can take account of the output signals to its own musculature is in a position to detect and factor out the decorrelat-ing effects of both moving objects and externally imposed body move-ment" (378).

More recent research examining discontinuities between prevalent models informing VE development and the real world supports Held's earlier findings. John Wann and Simon Rushton (1994) note that percep-tion in VEs is based on *deception*. It is technically impossible for an ob-server in a VE to maintain coherence across both perceptual domains of vision and "vestibular stimulation." Hence Wann and Rushton take issue with design approaches based on a belief that visual perception is capable of resolving on its own any incoherency between itself and motion perceptions that seem out of sync with it. They critique the limitations of current direct perception or ecological theories of sight (see discussion of J. J. Gibson hereafter) informing actual VEs and do so in part through the following thought experiment.

A subject on a swing dons an HMD through which she sees a textured virtual world. She is physically pushed from behind so that she feels motion, but she exerts no effort herself in moving. As a result, she experiences a sense of movement through the HMD. Although at a certain point her body begins to lose momentum, a sense of moving at the original velocity is maintained on the spatial display by which she interacts with the immersive virtual world. At this point, theories of direct perception on which VEs are modeled maintain that her visual cues would overwhelm the kinesthetic "vestibular" sense of balance, and that she would continue to experience the "ego-motion" of moving on the swing. Suppose, however, Wann and Rushton continue, that she is given a second push. Her amplitude increases. At the same time, the visual amplitude on view within the VE is decreased (1994, 339). Will the subject, whom direct perception theory had theorized in the first instance as continuing to perceive movement (because of the primacy of visual perception over all other sources), even though she had slowed down considerably, feel as if she is now slowing down when in fact she is accelerating? The authors are inconclusive in their answer, save to say that since, in addition to visual cues, vestibular effects are required within VEs to achieve verisimilitude, the resultant discordance between sensations interrupts the deception of self-motion. Technical lag produces "phase differences" leading to confused perception and physical discomfort. The deception of self-motion depends on *coordinating* visual and vestibular perceptive faculties. If this is disturbed, as current VE applications have the tendency to do—because they accord primacy to the visual and have underestimated the role of body motion within the visual process—the deception of self-motion is broken, and a return to the senses may express itself in nausea, vertigo, and disorientation.

Wann and Rushton point out that a vestibular sense of motion is neglected in much contemporary VR lab work. Yet there is a fair volume of literature examining effects of nausea in VEs (Biocca 1992c; DiZio and Lackner 1992; McCauley and Sharkey 1992; Pausch et al. 1992).[22] The approach of Dennis Proffitt and Mary Kaiser (1991), for example, would seem exactly the target of Wann and Rushton's critique. Proffitt and Kaiser well understand that motion is a necessary condition for visual perception, arguing that it is a sufficient condition for perceiving many environmental properties, but they do not think that motion need involve the user's body. Rather, reminiscent of the panorama's circular perimeter, which in sophisticated versions moved while the viewer remained stationary, they suggest that motion of representational icons can replace motility in VEs. It is somewhat ironic, recalling Tuan's comment that visual space is static and bounded, that VE designers Proffitt and Kaiser would target the inherent ambiguity of static displays as a limit to be overcome. They further claim that the entire dichotomy of figure-ground is an ambiguity rooted in this "static"ness (48), an observation that reads as an inadvertent admission of frustration with the limits of vision penned by researchers whose work is designed to further a reliance on visual mechanisms. Users "can always be made to have erroneous perceptions whenever they are constrained to view an object from a unique perspective" (49). The authors' solution to this undesirable deception is not to accord physical motility of users its due but rather to argue for additional visual cues such as familiar surfaces, increased "texture gradients," and the use of perspective to suggest depth or distance between figures and background surfaces. Integrating these gradients with representation of motion becomes key to their research. They argue that any 3-D object is capable of being perceived in a wholly satisfactory manner in two dimensions if its form is continually rotated, for according to "kinematic law," "objects do not distort when rotated and our perceptual systems were formed in the context of natural constraints. The exploitation of these constraints does not require that they be embodied" (52). There is the presumption here that the viewer is immobile, and that form is somehow deceptive; only movement counts. Further, there is a second assumption, rooted in the premise of individual access to this technology, that the viewer is alone, without the benefit of others with whom he or she might compare perceptions. For all the claims that VEs as an IT might constitute a kind of revised public sphere, by their comments on the perceptive limits of our visual apparatus, Proffitt and Kaiser suggest the opposite might be the case. Hannah Arendt (1958, 57) argues that any reality in a public realm relies on the "simultaneous presence of innumerable perspects." While a VE might offer an image of a public realm to its interactive participant-consumers, any such eventuality would need to pass the test Arendt establishes. "Only where things can be seen by many in a variety of aspects without changing their identity, so that those who are gathered around them know they see sameness in utter diversity, can worldly reality truly and reliably appear" (57). Proffitt, Kaiser, and Arendt likely all could agree that the common world is destroyed if the sameness of an object cannot be discerned. However, as Arendt notes, this sameness is discerned by our ability to share information with each other to fill in the gaps of any one individual's personal knowledge. It is not discerned by an increase in "texture gradients," or even in a physical ability to circle an object to confirm this sameness to ourselves. Yet in a virtual world, each user must be fully apprised of the reality of the representational object, precisely because confirmation from others will not be at hand.

Euclidean geometric laws apply independently of the substances that embody geometric relationships. "The laws of space operate as though space were unaffected by substances, as though space were empty" (Sack 1980, 62–63). For Proffitt and Kaiser, the adjective *visual* is assumed in their consideration of the ambiguity of static displays, or the lack of distortion in displays representing objects in continuous motion. It would seem that the visual and appearance of movement are all that really matter. The illusion of perfect clarity, to be obtained through the medium of immaterial light within a space modeled on Euclidean geometric law, demands elimination of ambiguity. Complete correspondence of image to reality, or of even the invisible to the visible, is the goal. The meaning of material substances matters less than how they and their images might both conform to a necessarily reductionist law. Substance is detached from the moral weight given kinematic laws in the same way that the Atomists detached the Void from the air.

I have been mildly critical of Jonas for the princely status he accords vision. However, I noted the connections he traces between touch, body motion, and sight. "The *motility of our body* generally. . . is already a factor in the very constitution of seeing and the seen world themselves, *much as this genesis is forgotten in the conscious result*" (1982, 152; second emphasis added). Facilitated by according primacy to sight and metaphors of vision, this forgetting (or, adapting a phrase from Walter Benjamin, a kind of "optical unconsciousness") is central to believing that VEs could wholly represent the entirety of our perceived world both as and in "disinterested," a priori, and thereby precausal images. Theories of direct perception, and by extension variations of empirically based correspondence theories on which much virtual technology research derives a philosophical and theoretical foundation, also stem from this forgetting. Such theories are not without challengers, but they predominate in American VE research and development.

Theories of Vision and Virtual Environments

> What is it like to see? Is it to take the thing seen as forming a totality outside oneself, and as being that in relation to which one orients oneself? Or is it to take the act of looking as the means to decide one's cognitive being in the world? (Harrison 1994, 231)

Western understanding of the nature of truth depends on the ability of statements to *reflect* accurately the entities to which they refer. Belief in symbolic representation as the best access to truth connects, in part, to radical Cartesian doubt as a proof of existence. Knowledge dispels this uncertainty and provides a measure of security from the "error" implicit in sensation. The Cartesian subject exists, in part, by asking "is it true?" This introduces a fetish for accurate representation, as it is this subject's mind that builds through thought a representation of the world. The input it receives *ought* to be as factual, and as direct a representation of external reality, as possible.

Outlining theories of perception and representation that underpin research in VR technology, Richard Coyne (1994) details how what he calls a "correspondence theory" of reality has gained the upper hand. Correspondence is driven by the desire to produce an ever more perfect copy of the "original" or external reality, whether this copy is a painting of a person or landscape, or a VE simulating the Gulf Stream corporate jet. Correspondence theory assumes that geometry and number, which are easily transposed to VEs, constitute basic foundations of the real world. As with Platonism—in which knowledge of "what is" comes through a life devoted to intellectual striving beginning with the study of mathematics, which directs the mind away from the senses toward contemplation of more "real" things (see Kitto 1964, 193–94)—the Image of perfect Forms is what is truly real and the ultimate goal. Mathematical patterns are more real than anything they might refer to in the physical world. Correspondence theory attributes priority to abstract entities as "vehicles" for seeking the truth. The reality of things is located less in their substance than in their stability, reliability, and intelligibility—"lawlike," ideal characteristics entirely consonant with what is possible in a VE. If geometry and number can be modeled in a computer system, then their representation as information will form an "accurate reconstruction of reality," a kind of Turing Test for photorealism and VR (and a triumph of representations of representations). This assumes that the world is entirely amenable to being computed, in part because an understanding of the lived world has been reduced or alienated to an "objective" world "out there" beyond the "wall" segregating it from the subject. The HMD used to access a VE becomes a "thinking cap" donned by humans assumed, under a merger of correspondence theory and the logic of symbolic representation, to require ever more perfect (two-dimensional) copies of reality to perceive the world.

Correspondence theory accords with the view of perception as a passive conduit for data/representations from the environment to the mind. To effect a sense of the real in VEs, ever greater inputs are needed. Just as the body is conceived as "an input device," virtual technology makes it seem that appropriation of reality is a matter of data processing, and given powerful enough computational devices, the "ultimate display"— Sutherland's (1965) perfect simulation of reality—will be achieved. To believe this is also to think that thought is only the manipulation of signs (Bolter 1991, 224). Although I am in these pages arguing against the world as reducible to being computed, the desire that drives belief that perfect simulation of reality might be achieved should not be dismissed as a quixotic aberration. At the least, such a dismissal forecloses the possibility of crafting plausible argumentation against such an undesirable eventuality, for it precludes or avoids coming to terms with why many people now desire such an (unreal) virtual state of affairs.

Discussing the turn from painting to photography, Stanley Cavell (1971, 20–21) finds technological differences less at issue than the metaphysics driving the trajectory toward greater "likeness" in Western representation. Cavell links this push toward likeness with "the human wish, intensifying in the West since the Reformation, to escape subjectivity and metaphysical isolation—a wish for the power to reach this world, having for so long tried, at last hopelessly, to manifest fidelity to another." More recently, in his history of Western painting, Bryson (1983) argues persuasively that from the Renaissance deployment of perspective forward, an ever present wish and push toward "the perfect copy" has motivated Western visual representations at least until the challenge posed by modernist art movements such as cubism and surrealism. Cubism and surrealism, and, for example, American abstract expressionism, were each successful in reflecting aspects of the social condition within which the movements were situated—disincorporating social relations, new ways of theorizing sight, multiple visions of society, technological alienation, and so forth, as well as the contextual genius of certain painters.

These movements, however, did not displace the push for a perfect copy Bryson identifies. The acceptance of realism by a purchasing twentieth-century public attests to this. Many more bourgeois living rooms feature mirrors, landscapes, and portraiture over the mantel than a Picasso or a Jackson Pollock. Although polyvalency in immersive VR in some ways extends cubism's 2-D work with multiple subjectivities and angles of vision, the push for a correspondence, particularly in American VR research, suggests the ongoing cultural purchase and celebration of correspondence as a theory and mode of representation.

The perfect copy (manifesting the quest for transcendence into an ideal utopian space, and the philosophical illusion that absolute clarity is available via "direct perception") also bears uncanny resemblance to the Pentagon's interest in VEs as providing the means to author "seamless simulations" that repeatedly and purposefully blur distinctions between what is real and what is not (Sterling 1993, 94–95). The Pentagon's goal reflects Umberto Eco's (1983) assertions that the gap that once existed under Platonism between image and thing has disappeared, so that cave shadows are increasingly hard to distinguish from the reality outside.

Belief that correspondence is the best theoretical engine to drive VR research is challenged by a *constructivist* orientation within the virtual research community (Coyne 1994, 66). This view, more in alignment with the empiricist philosophy of vision outlined hereafter, holds that human perceptions are always influenced by a culture, and that familiarity with the cultural assumptions built in to any one VE is a precondition for making sense of experiences had within it.

It should be noted that while the terms "correspondence" and "constructivist" have precise philosophical meanings for some, these terms are much more loosely interpreted by many virtual researchers. Even the use of these terms introduces certain difficulties. For example, Brian Gardner (1993, 105) understands constructivism as the traditional sense of perception that holds that light enters the retina and creates a complex mosaic from which the eye reconstructs the environment and objects in the world. In other words, his view of constructivism is similar to the Cartesian view of perception as inputs of sense data. Gardner's use of constructivism differs radically from Coyne's and is more akin to the latter's understanding of correspondence. Robert Schwartz's (1994) way of distinguishing between the two approaches Coyne identifies is helpful. Schwartz would situate correspondence theory as an innate or nativist approach to perception and would identify theories according a more central place to culture and memory as learned or empiricist.

Correspondence (innate) and constructivist (learned) approaches form part of the larger debate about visual perception that has sequentially engaged philosophy, psychology, and the cognitive and computer sciences for several hundred years. Between a conviction that direct perception can be obtained via visual data alone, and one that holds that perception is entirely socially constructed, lies a continuum of intermediate theories. Most virtual technology research locates itself closer to the direct perception or innate end of this continuum, though voices continue to struggle to be heard within this community about what is being lost by too narrow a focus on correspondence.

It is useful to outline the key distinctions between what Schwartz calls innate and learned arguments. Within the virtual research community, both are interpreted in visual terms. Yet the latter has greater ease in admitting the role of language and culture in organizing both the manufacture and successful perception of any necessarily culturally specific VE.

George Berkeley's *An Essay towards a New Theory of Vision* (1709) is echoed in subsequent arguments that the eye itself cannot stand alone as a mechanism of direct perception of spatial relationships. Opposing Berkeley's theory, James J. Gibson's more recent "ecological optics" or "ecological" approach suggests that the physiology of sight is in most ways sufficient for directly perceiving the environment around us. Gibson's theories have had great impact on virtual research, which downplays Berkeley's central thesis. However, aspects of the latter's work have not been entirely rejected. The "pesky" issue of vestibular disorientation and the spatial relationship between the user's body (which organizes an interplay of perceptive qualities despite correspondence theory's avoidance of this reality) and his or her disembodied point of view in a VE suggests that not everything can be resolved by resorting to ever greater visual realism and information density in spatial displays.

Berkeley's theory of vision is a precursor to psychological theories "of the perception of spatial properties of the world" (Schwartz 1994, 4). For Berkeley, distance cannot be immediately seen. The eye only sees the end point of distance on a line directed endwise to it regardless of the length of the line. Berkeley does not deny that we perceive distance visually, but he argues it is not *immediate* (8). What we identify as the perception of distance is not innate and must be learned. In contemporary terms, distance perception for Berkeley is both physiological and the product of social relations. Spatial perception in general and distance perception in particular depend not only on sight but also on its relationship to each individual's movement (9) and of the necessary learning how to do so that this implies: a position that anticipates to a considerable degree those of Held (1972), and Wann and Rushton (1994).

The stereoscope's invention in 1833 by Wheatstone (noted in chapter 2 as a precursive component to immersive virtual technologies) suggested that the discrepancy of visual cues each eye received was an important stimulus to depth perception. The stereoscope was seen to refute Berkeley's "one-point" monocular argument. Wheatstone demonstrated that the disparity of binocular vision offered a visual source of depth perception independent of bodily movement (Schwartz 1994, 40), though later research has shown that while significant, this *stereopsis* does not solely account for distance perception and is but one of several depth cues humans use (Gardner 1993, 107).

However, the phenomenal character of visual experience counted less for Berkeley than an ability to read experience for its significance. Fred Dretske (1969, 18–19) distinguishes between visual experience and significance, asserting that the physiology of sight differs from the seer's visual attentiveness to what is seen. For Dretske, sight visually differentiates a thing from its surroundings (26). This differentiation is a preintellectual, prediscursive capacity. Much of the confusion that attends the debate as to what constitutes perception is rooted in the language used to describe what we see and say we see (35) — an issue more fully taken up in the following chapter on space and language.

Berkeley claims that perception of spatial distance, magnitude, and situation is derivative, that visual experiences are signs for the *idea* of spatial distance, of magnitude, and of situation (Schwartz 1994, 85). For Berkeley, experience is constitutive of reality. For Searle (1995, 169), Berkeley's use of the term "idea" is analogous to the twentieth-century use of "sense data"; for Copleston (1994, 406), to clusters of material phenomena. Links between visual experience and these ideas are created over time primarily by movement and learning and, to a lesser extent, reasoning. To know the distance *x* means having an idea of locomotion, of body pacing, and the ability to reach out and touch *x*. We perceive distance on the basis of movement and signs or cues we have learned to correlate with distance (sometimes by habit formed through past associations), and which we interpret for significance (Schwartz 1994, 23, 85). In and of itself, visual experience has no spatial content if it does not also take place within a learning context integral to how we move through our world. This body motility, of which eye-hand coordination is a part, gives us "tangible ideas," and these furnish visual experience with its spa-

tial dimension. Spatial ideas, while partially depending on sight, also result from mental activity and depend on "geometric reasoning" (27). Thus
Berkeley is able to assert that the "notion of space is but an abstract idea
of extension"[23] — an assertion that tries to explain how space is conceptualized but reduces space perceived by vision to a sign or "idea" (Jammer 1969, 135). Later theorists working within "the Berkeleian tradition" have emphasized learning and psychic processing as key elements ·
informing what is held to be the derivative nature of visual space perception. Berkeley's original insight is reflected in Jonas's connection of
sight to body.

> The "possession" of a body of which the eyes are a part is indeed the pri-
> mal fact of our "spatiality": the body not merely as occupying a volume
> of space geometrically but as always interacting with the world physically,
> even when at rest ... the eyes alone [could] not supply the knowledge of
> space, notwithstanding the immanent extension of the visual field. (1982,
> 154)

The Role of J. J. Gibson's Theories in Virtual Research

The "ecological theories" of human perception researcher James J. Gibson
(1950, 1966) have heavily influenced the direction of virtual research[24]
(see Rheingold 1991, 143–44), though his positions on vision changed
over time, and proponents and critics alike interpret his theories in contradictory ways (Schwartz 1994, 125). Gibson asserts that there is vastly
more information available to us in light arrays than classical theories
of vision such as Berkeley's acknowledge. Mind-independent lawlike features of reality can be directly perceived through sight.

Gibson's assertions proceed from his study of human sight and perception of the *real* world. Virtual researchers have applied his arguments
about embodied direct perception to VEs, which are designed primarily
to facilitate experiences of a "distinct phenomenal visual quality." The
environment we are getting information about is a subset of the wider
world. VEs are designed to accord primacy to visual access. It would
seem that an adequate defense for applying Gibson's theories to virtual
worlds (applying theory to metaphor, or representation to representation) is felt to lie in the fact that both deal in theories of visual spatial
perception. Such a position forgets that the "lawlike correlates" Gibson
finds between light arrays and environmental properties in no way address
or have anything to do with the differences in material substance between
real and virtual worlds. The different kinds of appeal these substances

make to different aspects of our sensorium are each a part of how we make holistic sense of the world. To look at the world and say, "I get the picture," may be "lawful," but it is only a part of that making sense.

Comparing Berkeley's theory of vision to Gibson's suggests they approach the world with different understandings of mind. For Berkeley, thought, or what we now commonly think of as ideas, does not have a representational function. Ideas are akin to a screen between the mind and things. For Berkeley, "the problem of correspondence between ideas and things simply does not arise" (Copleston 1994, 436). Gibson's theory is more like a return to a medieval way of approaching the world. The mind is not entirely passive, but it conforms itself to objects. Gibson theorizes that we *directly* perceive aspects of the world through our eyes. This grasping involves an agency of mind. However, any agency present need not contradict this grasping as having been directly perceived in a systemic fashion, because for Gibson, what we grasp, and the order by which we do so, is predetermined by evolution (1966, 155). The decentering of subjectivity that such a conforming suggests is also what is sought in the virtual world, though again I would acknowledge that Gibson theorized a hardwired "reality" and not an ersatz variation on it.

This summary suggests in very crude terms something of the difference between Gibson's and Berkeley's theories of knowledge and how such theories inform beliefs, which then get transformed into technologies that influence social relations. The virtual scientific and engineering community's interest in Gibson is intriguing, given this community's focus on *representing* reality. Gibson finds sense data to be vivid. This quality of vividness is *reproduced* in a VE, but as an image of an idea of an original vivid impression. This doubly removed image of an idea is then offered back to users as an analogue for the vividness Gibson had directly perceived in the real world.

Gibson's followers also sidestep the substantial agreement between Berkeley and Gibson that sight's main role is to give us information about our environment, and *not* to offer experiences having "distinct phenomenal visual qualities" (1966, 126). Both men privilege the productive faculty of vision over any aesthetic consumption that it might abet. An active sense of sight is action's ready guide. Correspondence and constructivist theories share a belief in technology as the driving metaphor for nature. Coyne notes that VR research constructs everything constructed according to a "productionist metaphysic." "It is as if to say: 'nature is constructed, so let us re-construct it in a computer'" (Coyne 1994, 68).

Early in his career, Gibson (1950, chap. 3) distinguished between an experience of the *visual field* and one of the *visual world*. Both depend on light. However, the visual field has boundaries and is experienced when we focus on the qualitative aspects of the appearances of things. This experience accords with the physiological properties of the retinal image transmitted to the brain. Railway tracks appear to converge because their images first do so on the retina. What we then do with this physiological experience of the visual field is to experience a visual world. This higher-order experience is similar to the notion of perception. The visual world has no boundaries. It is continuous and panoramic and "surrounds us for the full 360°" (28); we discover our world by sight, even though not every aspect of it is available to us at each moment we are seeing. Gibson suggests that an experience of the visual world is not mediated by judgment or reason, and that we see much more clearly than we realize. Information contained in visual stimuli is dense enough to survive the transition from the three-dimensional world to the two-dimensional retinal surface and still represent spatial distance. All the information needed for perception is directly present in the structure of light reflected from objects and events in space. The observer (and here one may perceive why VE engineers find Gibson's theories so resonant) is "immersed, drenched, in this information . . . [it] specifies the relationship of things one to another" (Campbell 1982, 203–4).

Gibson's *ground theory* holds that we directly perceive distance, in part, by detecting the difference between the fine *texture gradient* (1950, 78–80) of, say, a carpet five feet away, and the more muted texture gradient of, say, a painted clapboard wall seen from thirty feet. Texture gradient, seen at a distance, is a spatial property of environment and not a cue or a sign we subsequently use in mediating or understanding what it was we first saw. We see size as the distance between two points, and distance as the size between two points. Although Gibson (1966) conflates Euclidean geometric measure with rigid "terrestrial space," at the scale of individual experience, he would also seem to embrace a relational conception of space, but in a way that accords agency to distance as visually perceived. "Our hypothesis is that *the basis of the so-called perception of space is the projection of its objects and elements as an image, and the consequent gradual change of size and density in the image as the objects and elements recede from the observer*" (78).

For Gibson, texture gradients of things in themselves are an aspect of perceptual invariants, which remain the same over long periods of time.

Perceptual invariants owe a debt to Plato: they are theorized as perfect, unchanging forms. However, they are not beyond the reach of our senses. Because texture gradients are invariant, sight has become attuned to their direct perception. Perceptual invariants and texture gradients seem updated conceptualizations of Hume's vivid impressions. Prima facie knowledge of the world by human "perceptive systems" is objective and meaningful knowledge (see 1950, 198–99), and as it is "given" by evolutionary requirements, the theory works to place itself above critique. Arguably, the theory minimizes human intentionality vis-à-vis technology — an irony given the *interactive* nature of VR — and it also helps maintain the standard socialization model that posits humans as blank sheets of paper upon which are impressed values and knowledge.

Nevertheless, Gibson has a synthesizing grasp of the power that resides in images. "The spontaneous activities of looking, listening, and touching, together with the satisfactions of noticing, can proceed with or without language.... Observing is thus not necessarily coerced by linguistic labeling" (1966, 282). His theories support an understanding that images offer a means of communication operating very differently from language and texts. They suggest that to speak of either "picture languages" or "reading an image" is a misleading language game and an obfuscation of images' power. Not every semantic can be placed under the umbrella of language (see Langer 1985, 100).

As theories, perceptual invariants and texture gradients are innate and may be understood as restraints that direct human creativity and call attention to the limits of human possibilities and human consciousness in determining the world. Yet engineers applying Gibson's theories to VEs would appear to feel no such constraint. Indeed, the opposite seems the case. Texture gradients are seen as a way to overcome the vast storage and computational difficulties in modeling geometric form in VEs, as witnessed by work at the University of North Carolina at Chapel Hill, the premise of which is contained in the name of an ongoing project in the Department of Computer Science: *Virtual Backdrops: Replacing Geometry with Textures* (1997). Unlike Gibson's perceptual invariants, VEs are more like a latter-day visual Esperanto for the eye, whose engineers believe they can avoid the pitfalls that have faced all earlier utopias premised as spaces from which to access absolute clarity. Incorporating Gibson's ideas into an act of personal contemplation is one thing. Such an act is like the classical understanding of communications as a closed *system*. It is a different matter, however, to apply his theories to a conception of the world intended to operate within an information *network* that is rather less closed. Engineers are theorizing and building virtual technologies motivated by a belief that they are merely "transcribing" a universal truth directly from the environment, because (by extension) they too are hardwired and therefore, as conduits, could not do otherwise. Any lesson in limits Gibson might offer is thereby made to justify or mask a failure by research workers to fully think through the implications of the theories of knowledge they misapply.

Note that Gibson is concerned with how we see the *physical* world. Dretske's observation about the difficulty that resides in the language used to describe perception is important. Gibson is writing in the language of facts, and not the language of states (Dretske 1969, 165). He is not speaking about "seeing" anger or "seeing" a quality of inner emotion. To say I "see" your anger (which Gibson does not) is not the same as to say I see your face and I will always recognize it as your face because of light's transmission of perceptual invariants (which he does). Some of Gibson's current disciples now argue that emotions are also associated with perceptual invariants. Dretske's valuable distinction—echoed by Searle (1995)—is set aside. Gardner (1993, 111) considers that "emotions, behaviors, and personal actions may be viewed by others as social affordances [that] may be visually communicated through their associated perceptual invariants." Now, Gibson coined the term "affordance" as a substitute for *values* in order to shed the "old burden of philosophical meaning" associated with the latter (1966, 285). Like an actor on a stage, a character within a VE will represent the perceptual invariants that communicate (or in Gibsonian terms "furnish") the essence of the character portrayed. The difficulty with this analogy is that a stage actor communicates with his or her whole being. Research and development of VE spatial displays is based on a belief in perceptual invariants and "essences" that seems more akin to psycho-physical systems of explanation such as Renaissance physiognomy, or phrenology, or metaposcopy (face reading) than the stage.

The iconographic "language" on view within VEs, as Susan Buck-Morss (1989) notes in her superb discussion of emblems, is inherently unstable in meaning. Although the output of a polygraph, say, might represent an emotional state, emotions are not empirically verifiable in and of themselves. To suppose that anger can be seen and modeled with clarity assumes that anger—as opposed to the person who is angry—looks some way to us. This supposition confuses a person's state with a thing to which he or she is related. "It is a mistake to take observable behavior for psychical truth" (Burgin 1989, 23). Belief in a postsymbolic communication via VEs seems premised on this misconstrual, and anchored in an implicit magical belief in "future foretold."

For Gardner, in a VE, the *quality* of what is to be represented is key, and "what defines quality [for interacting within a VE] are these environmental invariants, which drive the human perceptual system" (1993, 106). Therefore the type and level of detail to be represented in any VE should depend on whether or not the detail is of a perceptual invariant. Gardner and others who apply Gibson's theories to VEs blur the distinction between what it is to see and what it is to represent this seeing— what W. J. T. Mitchell (1994, 8), in the slightly different context of referring to landscapes, calls the blurring of "the distinction between viewing and painting." Gardner transposes Gibson's work on the psychology of vision to VEs, which become an *expression* of what is seen. I have argued for understanding VR as both technology and medium. However, influenced by a McLuhanesque notion of the technological extensibility of the self, what Gardner seems to propose is that VR as a new medium and new "nature" is also a new human sense.

I am not objecting to this so much because it represents a psychasthenic blurring of the boundary between human and nonhuman. The leaky boundary of our own bodies, which exist along a sliding continuum between nature and culture, already confirms something of this hybridlike indistinction that straddles the discrete realms of nature and culture. What *is* problematic is to have used language—the language of Gibson's texts, the software that underwrites "texture gradients," the written description and theory that Gardner provides—in short, to have recourse to mediation in order to suggest that something more *pure* than this very mediation might now be on offer through a new media form. It is to say that users of a VE might ignore the language and mediation base and act as though they were in direct receipt of a sixth sense somehow beyond culture's "contaminating touch." To be fair, Gardner is promoting VEs within a confusing ideological arena in which slogans of individual primacy advance acceptance of these technologies, which are precisely about relaxing distinctions between self and world. In other words, VEs are about mixing up, by intense immersion in mediation, the old binaries of self and other, and of nature and technology as a cultural representation of nature. Promoters of the technology operating within the hard sciences seem to have as much difficulty acknowledging this as do critics of VEs' effects writing from a humanities perspective or the academic Left.

"A perceiver is a *self-tuning* system" (J. Gibson 1966, 271). The theory of texture gradient is an exceedingly technologized and geometric view of how humans sense and perceive their world. In many ways, Gibson's use of the concept "element" seems to update Greek Atomist and neo-Pythagorean belief in the prime reality of number, along with Descartes's assertion that number is the basis for understanding space. Gibson posits that the number of texture elements projected into a specific area increases as one looks at the ground from farther and farther away. An invariant ratio is argued to exist between the number of texture elements and their distance from the eye. When the distance from a texture element is, say, doubled, its projected height is reduced by half. Therefore a continuous gradient of ground texture is always available for distance perception (Gardner 1993, 107–8). As I walk away from an object, it appears to shrink, as do its surroundings. However, the object "still occupies the same number of units of space, but those units all shrink together as they become more distant . . . the relationship between the [object] and its own space remains the same [to visual perception]" (Campbell 1982, 205).

Although highly contestable, Gibson's assertions that we are hardwired for number and geometry, and that this forms the basis for how we make perceptual and then rational sense of the world, pale in comparison to how the theory is now applied to rationalize both an ascription of magical properties to space and the human body's relationship to movement depicted within VEs. Gardner argues that in the real world, "when you move forward, all the texture flows away from the point toward which you are moving. This directly perceived stable point and *texture flow* patterns in the visual field enables [sic] extremely accurate goal-based navigation" (1993, 108). Gardner connects this flow of texture to a second phenomenon in which closer objects appear to move faster than those farther away. Simulation of these different speeds of movement is a goal of virtual engineering. Interestingly, when Jeremy Campbell — also an advocate of Gibson's theories — seeks to describe this effect in the real world, he employs a framing device and describes our perception of the different speeds at which closer or farther objects appear to move as if "seen through a train window" (1982, 204).

My own mundane experience of movement does not confirm Gardner's assertion. My body motions produce messages that interact, interfere, confirm, and contradict to varying degrees and at varying times with whatever direct perception I might potentially be receiving from texture gradients. However, my own experience with VEs also confirms that the virtual world operates very much the way Gardner describes. Simple applications may not require a high degree of texture, but the most sophisticated commercial and military applications currently on display are highly textured, highly luminous, and vivid. Wall surfaces, for example, are often wallpapered and multicolored rather than plain. This makes for easier spatial identification of 3-D things represented in a 2-D spatial array. The user is less likely, therefore, to "walk" into, merge with, or get lost in such nether "spaces" as the interior of a wall, which could not be accessed in the real world. Gradient seems to be understood by VE designers as "grain" and "interior illumination." Tabletops and cabinetry are grained. Water and flame shine with an inner "radiosity." Interior walls, or the streets of the abandoned village in the VE described in the introduction, appear to flow by one's sight as one navigates or flies through the VE. The atomized nature of "elemental" texture gradients transposes perfectly to the pixelated realm of spatial displays; Gibson's theory has become an aspect of an entirely spatialized reality. The application of flow speeds causes close-up objects to rush by compared to the slowness with which objects farther afield seem to move. Campbell's train window merges into the HMD and the spatial display it reveals.

If VE promoters truly seek spatial displays that offer an access window onto the philosophical quest for absolute or perfect clarity, then they will need to better consider that in applying Gibson's theory of texture gradient to the virtual world, the body motility he identifies as interacting with such gradients must operate as more than only a mechanistic device for moving the "grasping" eye through space or moving the space in front of the "grasping" eye. When VE theorists speak of inclusion of body motion in VEs, they often reduce the meaning of our bodies to something akin to a dolly on wheels. Gibson, however, distinguishes between "imposed proprioception" — occurring when parts of the body are passively transported, and the eyes are stimulated by motion perspective without the participation of the muscles (Held's experiment with the kittens) — and "obtained proprioception," which occurs when one performs an action with any of the human body's motor "systems" (1966, 45). Theorists such as Gardner conflate Gibson's distinctions into a retheorized "imposed proprioception" that conceptualizes vision, and an image of movement in representational space, as capable of standing in entirely for the whole of bodily motility. It could be argued that Gibson provides some justification for doing so when he states that looking at scenery is not a passive state, as if asleep, but an attentive, contemplative, nonperformative activity (45). In contemporary VEs, motility, therefore, is not denied, but made an adjunct to sight's direct perception of invariant texture gradients. When one puts on the HMD, one first orients oneself in space with respect to the direction in which one sees. As in the real world, when one turns one's head in a VE, one sees a changing field of view. To travel in a VE, one first turns one's head in the direction one wishes to go. Activating a handheld control device, one then "flies" in this direction. Vision is paramount in determining orientation. One flies as if a bird, yet although our eyes perceive a horizontal range of 180 degrees, most HMDs contain a viewfield of only 105 degrees or less, instantiating, I believe, a quality of spatial lag that is underaddressed in the current literature. Peripheral vision is not really present, partially because a vestige of framing remains.

The following experience was recounted to me by several "novice" users I questioned following their initial immersion in a VE. They begin their experience and orient themselves visually within the spatial display. Then they turn toward the left or the right in order to view other parts of the environment. However, turning back to the direction from which they believed they started, they discover they are not in the same "place." It was explained to me by individuals very conversant with both the hardware and software in question that the users had moved their entire bodies slightly, and not just their heads. This caused the position tracking device to adjust to their movement and to relocate their points of view in the display, suggesting that to not become "lost" in a VE will place a disciplining demand on many individuals to become more self-conscious of a "sequential" relationship between vision and head versus whole-body motility than is required in real life. Additionally, the technology permits users to see like birds or pilots in flight — up, down, sideways, and all of this in rapid order. Given Gibson's reliance on evolution as a defense of texture gradients, it is ironic that a VE engineer indebted to his theory may confuse users precisely because they have not been "hardwired" by evolution to fly like birds. Finally, at the outset of a VR experience, a sense of the location of one's virtual body — most often represented by an icon of the right hand — corresponds experientially to where the vision point seems to be. However, it is possible quickly to extend the vision point far beyond the virtual body's location, fracturing a corporeal understanding of eye-hand spatial relationship. It feels as if one's eyes are attached to a very extensible and flexible coaxial cable that moves in the "six degrees of freedom" all at once. The user must learn to discipline and control movement ratios between hand and helmet motions when manipulating virtual objects and moving through virtual space. Otherwise the hand may disappear from view and "get lost." In the real world, we do not depend solely on vision to update information on which eye-hand coordination is based. Although the sense of immersion within a VE can feel quite convincing, such experiential anomalies — to which one can habituate — can make the experience feel less "true," or even unreal, and demand a bodily conforming of human abilities to the capabilities of the machine.

It is not that aspects of VEs might not be true. They have existential and social reality and empirically verifiable effects. However, VEs occur in and reflect a culture that believes seeing equals knowing, and one that accords representational correspondences the status of a privileged access to truth. An immersive image technology offering the promise of access to truth trades on the power of image to suggest that the social intentions behind the technology precede or are immune to causal involvement. If image is accorded the status of *being* (Jonas 1982), it becomes an "essence" that seems above or before the contestatory "narrative fray of becoming" and the sphere of human agency based on situated knowledge. The use of Gibson's ground theory and its absolute notions of perceptual invariants deflects consideration not only from how visual spatial perception might be culturally and linguistically inflected but also from the fact that not everything can be pictured. As some realities are inaudible, so too are some invisible. The increasing cultural primacy of all things visual furthers a belief that the situated and pragmatic knowledge behind virtual technologies is aligned with a universal truth reified by the seemingly a priori nature of images. This helps ensure the technology's reception as an unmitigated social good. Illusions of perfect clarity within VEs are achieved by forgetting that laws linking geometry and number to the real world are made to work in metaphoric ways that do not speak to causal effects engendered by the different substances composing the different realities on either side of the interface between virtuality and reality.

Claims for postsymbolic communication between people assume we could circumvent our need to use symbolic language forms as part of interpersonal communication. Postsymbolic advocates who have been inspired by Gibson's theories of the natural environment now focus on the perceived limitations imposed by language and abstraction on interpersonal communication. Conscious awareness of the formalism and necessary theatrical component of public communication is somehow always suspect; everything must "flow" as effortlessly as between the intimately acquainted on a good day. Gibson, however, was concerned with how we receive information from our environment. Possibly this concern could be restated in terms of how the nonhuman world "communicates itself" to us. Although I am critical of how Gibson's overreliance on texture gradient can be used to minimize the qualitative and hence fluid roles of memory and acculturation that Berkeley's theory recognizes, Gibson is not writing about any supposed ability on sight's part to communicate any and all "emotional states."

In a sense, Gibson and Berkeley are each right and wrong. Berkeley is correct in having noted that sight is not an "independent agent." Even Gibson's disciples acknowledge the functional contribution of motility to sight, though they often directly proceed to confuse body motility with represented motion. However, Berkeley may overstate his case in asserting that sight is always secondary to movement. The reverse applies to Gibson. His sophisticated grasp of the relationship between distance and size as perceived by sight is not matched by his mechanistic assertions, under the rubric of "ecology," that sight is invariant (and hence immune from cultural influence or "education" through touch or eye-hand coordination). His belief in invariance is part of a wider cultural yearning that absolute clarity might be achieved via the utopian dream of fully unambiguous communications based principally on images. Although Gibson is writing about sight, his findings have been applied to the wider understanding of vision as representation. Berkeley's incorporation of learning and motility positions his work more closely to the meaning of vision than sight. However, human perceptions are informed by more than sight, hearing, touch, body movement, taste, and smell, each somehow held apart from the others in abstraction and splendid isolation. Our perceptive faculties are more integrated than this — eyes included — and are fed by multiple sensations received cotemporally. At times this means a synthesis of taste and hearing, at others sight and touch. Like the centered and circular space of ancient Hellenistic thought, the possibilities are finite and boundless.

5. Space, Language, and Metaphor

Neil Smith and Cindi Katz (1993) have criticized the widespread "rediscovery of space" by social theorists operating across a spectrum of academic disciplines. What has been rediscovered, in fact, are unstable metaphors of space. In the rush to city-as-text and "spaces" of power, real exploitation in real cities and other working environments is exacerbated by the impolitic use of metaphor, which causes concepts to be isolated from the active world to which they refer. Insufficient attention to the history and production of conceptions of space has kept theorists from noticing the metaphysical and political wrap that surrounds Western *conceptions* of absolute space. As a naturalized metaphor, absolute space has been assumed by these theorists as a given, as if the category of thought it represents were of the same substance as trees, rocks, or the tides. Absolute, geometric space assumes an empty field onto which humans can place discrete and mutually exclusive locations as points on a grid. Such a conceptual overlay is crucial for the spatial mobility of capitalist production and accumulation activities that increasingly depend upon the political disarticulation of one place from another. The social relations that theorists had hoped to criticize by the use of spatial metaphors are actually further strengthened by this ill-considered academic strategy, given that the historically specific conceptual basis of the "space" under contract as metaphor is ignored (ibid.).

Smith and Katz's point is well taken, yet only to a degree. Ill-considered or excessive use of metaphor, for example, influences the cultural understandings under which the "metaphoric" virtual technologies I am examining get imagined, built, and used. "Cyberspace," "Virtual Reality," and "the Net" are obvious examples. The moral limits to metaphor also express a disciplinary wariness of the power of linguistic metaphors to facilitate *real* communication between people, *and* to act as vehicles for conceptualizing movement and communication across real environments and imaginary spaces. Yet despite the limitations of metaphors — of which any theorist has an obligation to remain aware — there is no space worth *discussing* that somehow might be understood or represented apart from language and the necessary use of metaphors in human acts of communication. A "perfect" division between the material world on the one hand (and the technology that is a part of this world), and the give-and-take between language speakers on the other, is not "for real" and, indeed, would seem ironically idealized. I agree with Donald Davidson, who makes the point:

> There is no limit to what a metaphor calls to our attention.... How many facts or propositions are conveyed by a photograph? None, an infinity, or one great unstable fact?... A picture is not worth a thousand words... words are the wrong currency to exchange for a picture... the attempt to give literal expression to the content of the metaphor is simply misguided. (Davidson 1978, 45)

Metaphors of reality (of which a VE is an excellent example) are unavoidable, hence the vigilance required to distinguish them from reality, and to keep them from being frozen into monuments or idols (Jones 1982, 5). The resulting idolization represents a failure to recognize human authorship of conceptions of reality. Metaphor is one of our most valuable intellectual tools, yet it is also at the heart of the apparent dichotomy between reality and consciousness. It offers an awareness of duality within sensation, yet also the pretense that separate things, linked by metaphor, are one. We agree to a tacit "as if" (ix), exemplified, I would add, by the "consensual hallucination" of our participation in a VE.

Metaphors act as discursive mediators that allow us to conceive and maintain an ongoing relationship with the natural world. This connects to the mimetic, even magical quality of language as a practice whereby objects' expressive elements are brought to speech (Buck-Morss 1993, 322). Metaphors, however, do not transfer understanding from one sphere or thing to another along the lines of an empty conduit. Rather, they initiate a three-part dynamic among themselves and the two disparate things they link. The power of metaphors resides in their assumed cultural bias and ability to inflect meanings through the associations they create between themselves and one thing, and then between themselves and the second thing being associated with the first.

We are encouraged to think of ourselves as fluid, emergent, decentralized, multiplicitous, flexible, and ever in process. The metaphors travel freely among computer science, psychology, children's games, cultural studies, artificial intelligence, literary criticism, advertising, molecular biology, self-help, and artificial life. They reach deep into the popular culture. The ability of the Internet to change popular understandings of identity is heightened by the presence of these metaphors. (Turkle 1995, 264)

Assessments of virtual technologies often stress their links to science, whereas considerations of the VEs these technologies make possible often seem overly focused on theorizing the subject "therein" as an endless and pleasurable play of secondary effects. Both approaches neglect the bridge across this gulf of understanding between science and cultural studies that VR has already forged. VR is a hybrid of language and technology. I am critical of the ways in which it draws language and technology together, partly because at the same time it masks this joining. However, Smith and Katz seek a purity that metaphors and mediation are precisely not intended to supply. Need purity and mediation always be theorized as oppositional to each other?

Our participation in the social imaginary is intimately connected to the mutable power of metaphor. Because the meaning of any one metaphor is impossible to freeze, there is a tendency to encapsulate the metaphor's power in a metaphysical wrap by those with a stake in maintaining specific cultural associations among particularly important metaphors and the concrete and ideal realities they link. This crystallization creates the illusion that the metaphorics in question — previously communicatory mediations between disparate things and processes — are the same as absolute and eternal truth. At this point, icons become idols. Such a metaphysical move relies on a failure to distinguish between what we see and how we represent this seeing — between "observable behavior and psychical truth" (Burgin 1989, 23). Although the attempt may be made to conflate sight and metaphor, the two are not the same, regardless of the synergy released in combining the powers of both in optical technology.

Images arguably offer a more immediate communication than text, but this is possible only if and when enough potential viewers already share common understandings as to what constitutes a core of meanings attached to the visual symbols. Images are metaphors for these meanings, which have been negotiated through language (though the resiliency of these meanings is often revealed by the relative cultural durability of images). Nevertheless, the link between meaning and image is not eternal and is without guarantee. Because images are arguably more directly received than abstract text, they are subject to a wider range of individual interpretations; their meanings are policed (through the use of text), fought over, and subject to change. To ignore the power of metaphors, however, would be to take all images as literal expressions (Bal 1994, 193). Much of the hype surrounding VR ironically fosters this idealism by suggesting implicitly that we are "hardwired" to directly perceive reality through vision, and that this new form of technologically enhanced vision will lead to the promised land of "postsymbolic communication" (Lanier, in Biocca and Lanier 1992, 160–61).

Two additional points should be noted. First, new forms of icons are sometimes required to elicit new instances of metaphysical belief. Second, a belief that "purity" or a separation might be retained between language or culture on the one hand and individual (sighted) perception on the other is a fiction. To consider a virtual environment from the perspective of sight in isolation from the hermeneutic of language, or vice versa, ignores how sight and language interdepend. An epistemology constructed only along the lines of one or the other prevents coming to terms with how the hybrid potential of metaphor joins the two.

The West's understanding of the relationship among vision, sight, and light has been shaped by various contradictory and complementary concepts often expressed through metaphors of light. The interplay among vision, sight, and light constitutes the essence of experience within a virtual environment. Yet the history of optics and vision is told in language. This is why metaphors of light are discussed here, and not in the previous chapter on vision. To employ a spatial metaphor, in a VE, vision sits "atop" language in a kind of upstairs-downstairs relationship. The images within a VE are translations of the grid of ##### and text that are "residents" of the invisible but organizing language domain running "below" the display. Although vision and language play different, complementary roles in virtual technology and VEs, the invisibility of the language "base" of code and software on which the iconography depends leads to a blurring of the meaning of their different roles. This is reflected in the phrase "picture language" — a blurring of "stage" and "script" — which is used to describe the images or content of the VE.

In this chapter, I examine how light metaphors (re)position spatial relationships between seeker/viewer/subject and light as a source of truth, and how these metaphors inform VR. Aspects of specific connections suggested in early Hellenistic light metaphors among space, light, and the subject retain ongoing salience. Virtual technologies draw physical sight and metaphors of vision together. In so doing, they participate in a metaphysics of light as old as Plato's cave. The review that follows traces the evolution of three metaphors of light.[1] The first metaphor situates seers as being *in* the light that shines on high. In the second — after light is made to recede conceptually from the earth as its home — humans are no longer in the light but look *into* it at a distance. Finally, the modern subject is both *in* and *of* the light; in addition to looking in to the light, a separate inner light is posited as illuminating the individual's rational search for enlightenment.

Light Metaphors and Virtual Technology

In the Light

Light and its association with the day have been central to many cultures' metaphors of transcendence, the good, truth, and power. The early Greek philosopher Parmenides believed that darkness was overcome in the essence of light. The concept of light originates in the primordial view of the world as darkness and light. Enmity between these forces generates awareness that nothing is self-evident, including truth. This does not mean that the dark is denied its due. Everything has a place in ancient Hellenistic expectations (Walter 1988, 185), the dark included. In the essence of light, darkness is overcome, and intellectuality surmounts material actuality. Light is the "wherein" of nature and not a component part. Light is visible only when reflected by objects and is transcendent because it is not *of* the matter it reveals. Rather, like space, light articulates relations between this and that, here and there. Early classical thought understood humans as being *in* the light. In a similar way, we may think of ourselves as objects arrayed *in* space, a space in and by which we relate to other people and things.[2]

Transcendent light also bears conceptual similarity to communication understood as establishing a spatial relationship between copresent, coexistent senders and receivers. In chapter 2, I referred to Aristotle's observation that the power of sight both makes us know *and* sheds light on the differences between things. Privileging sight also privileges difference: the break between our self and the world around us affords the best means available for accessing knowledge. Yet for Aristotle, the reality of the seer and the reality of the seen are illuminated identically. Both are in the light. Seeing and light connect the seer and seen in a dialectical relationship. Darkness does not possess this discriminating power to reveal the position of objects in relation to one another. "Aristotle takes seeing to be not primarily some occurrence in the subject, but rather the visible's showing itself."[3] Both what is rendered visible by light and light itself remain external to individuals, who distinguish among themselves, light, and other objects that light illuminates. This distinction between self and the world-as-viewed helps orient us in the world.

Plato's cave metaphor, however, transforms light into an *idea* of the good. For the ancients, light, which gives all else visibility, does not have the character of an object. Platonism accords to light an implicit metaphysics because light is materially different from what it reveals to the eye. Ironically perhaps, "a way of expressing the naturalness of truth turns into its opposite: truth becomes 'localized' in transcendency" (Blumenberg 1993, 33). Light becomes a metaphysical truth, and partially because of this, light, along with the truth it carries, is conceptually withdrawn from the *kosmos* or world. Furthermore, despite Plato's identification of the "eye of the soul" and the "light of reason," and Aristotle's connections among vision, desire for knowledge, and sensual delight, no Greek thinker really explained which *material* properties of sight might qualify it for such "supreme philosophical honors" (Jonas 1982, 135). Plato, writing about vision, is most often using a metaphor of insight or pathway to knowledge and enlightenment. He is not implicating *sight per se,* save to the degree that his use of visual metaphors trades on mechanisms of seeing.

Parmenides' influence on Plato is considerable, and Plato's cave allegory does not deny the existence of dark places so much as suggest the natural connection between Being, light, and truth. The cave is a place metaphor for the *kosmos.* It is also a "doctrine of the restriction of human knowledge imposed by the body," which does not allow us to grasp truth, but only shadows and echoes (Couch and Geer 1961, 496). People trapped in the cave learn to love the illusions "projected on the walls of the dungeon of the flesh" (Heim 1993, 88), and it is also here that light is seized, exhausted, and *lost.* Freed of the temptations of this limited earthly realm, those formerly trapped in the cave can ascend to the realm of active thought. However, few mortals are equal to this task despite the classical imperative that being in touch with God or the "Idea of the Good ... was essential to full being" (Taylor 1994, 28).[4] This moral conundrum provides a second reason why light is detached from the earthly realm and metaphorized into salvation and immortality. Further, "light, now otherworldly and pure ... demands extraordinary, ecstatic attention, in which fulfilling contact and repellent dazzling become one" (Blumenberg 1993, 34). With the cave metaphor, light is already withdrawn — in a kind of "cosmic flight" — from a connection with (human) nature to a more supernatural realm. Any former prisoner of the cave who might ascend toward the pure light would look back with compassion on those left in ignorance below. As if prefiguring the twentieth-century dynamic of endless circulation within digital realms of information, such an illuminated, cosmopolitan individual would never return to the cave, or a life among the (embodied) shadows, even though complete wisdom or virtue would forever elude her or his grasp (see Kitto 1964, 498–99).

Cicero (106–43 B.C.) made Greek insight available to Roman culture with his translations of Greek philosophers. Amalgamating different theories of light, Cicero developed the concept of "natural light," linking "the metaphor of light with inner moral self-evidence" (Blumenberg 1993, 35), thereby somewhat reorienting a metaphysics of light. This *naturalis lux* would eventually filter down to inform Enlightenment assertions that humans also constitute a light source by virtue of access to this inner light as a gestating source for the self. In earlier Greek thought, light articulated a universal space in which all were illuminated equally. Cicero, however, conceives of human life as existing in a clearing that light makes for our occupation. It shines in an "economizing" fashion with respect to the space it illuminates, even though this clearing is a "dazzling envelope...pure and absolute" (36). Darkness is beyond the clearing, a "natural background zone."

Greek *theoria* attempted to comprehend the divine. Socrates, for example, understands *logoi*,[5] or an idea transmitted by speech, as facilitating observation of the truth of beings, given the mutual flight of light and truth from the world. By the time of Cicero, *logos* as a more discursive material power has fallen from grace, and the mind must be "redirected to the ineffable and nonconceptual contemplation of pure light" (Jonas 1982, 141). The "word" is judged imperfect and may lead one astray from the original or absolute to which it refers. To *look* is to know, though to Cicero this looking may refer to an interior examination performed in the clearing that light has made. Cicero "downsizes" the ancient space that light illuminates from on high to one more in keeping with the finite spatial requirements within which an advanced (Roman) culture might take place. Not only are the good and the moral at the center of this discrete clearing, which is visible and illuminated from above, but a second "internalized" light begins to emanate from within and takes the moral and aesthetic form of *virtu*. If the battle between light and dark suggested to Parmenides that nothing was self-evident, Cicero's repositioning of light begins an evolutionary process culminating in the notion that light illuminates being "present to oneself." The self—a metaphor for light, and a source of this metaphor—starts to establish a moral claim on determining what might be true. With VR, an appeal is made to self-illumination to augment being "present to oneself." Exterior light will enhance the truth of interior subjectivity, though this introduces the ironic risk of a remedievalization or reexteriorization of "consciousness."

In Greek thought, it is possible to really know only by light-dependent vision. However, as a result of reconceptualizing the spatial relationship between mundane humanity and light so that the latter comes to be thought of as emanating from a distinctly nonearthly realm, the secondary idea develops that the Good light self-squanders along its "downward" path to illuminate something or someone already fallen from grace. Light comes to represent a loss of self, or self-humiliation. Therefore, the post-Roman Neoplatonic Christian worldview has difficulty in accepting the now scandalous suggestion that the cosmos might represent a mistake or decline of divine light, or that good and evil might share the same root (Blumenberg 1993, 41)—as was the case in the earliest Hellenistic philosophy.

When Christian thinking reworks Greco-Roman light metaphors, it introduces a distinction between light a priori to earthly beings and *created* by God on the first day, and the multitude of earthly lights. In Exodus 3:4, God appears to Moses in a burning bush. The Bible "uses the element of light as the medium in which God becomes visible to man" (Jammer 1969, 36). The New Testament explicitly identifies God with light. In John 8:12 it is written "Ego sum lux mundi." God now becomes the reference source *behind* the light that emanates from his divine will. This distancing, which makes light a thing *and* a symbol, accords somewhat with a Neoplatonic positioning of seers as looking into a separate light, the source of which is now withdrawn to on high. The logical conflict between a Christian insistence on light versus evil and the earlier classically inflected understanding is relaxed with the return of some of light's metaphysical powers to a God-as-origin. A similar dynamic is at work in Jewish cabalist thought. Originally, light occupied the entire universe, but the Holy One withdrew his light and "concentrated it on his own substance, thereby creating empty space" (Jammer 1969, 37). Both accounts relocate light's agency away from the world and to a transcendent power.

Augustine's (A.D. 354–430) conversion to Christianity was facilitated by his reading of Platonic philosophy. In Plato he found a basis for believing in and understanding immaterial entities, forms, and ideas, which in turn permitted him to internalize the logic of Christian theism (Reese 1980, 39). Augustine redirects Christian theorization of light back to the classical seeing *in* the light, yet he also conceives of an inner light that is "behind" the self, a spatial move that returns the notion of origin or "wherein" to light but also renders looking *into* it impossible (Blumenberg 1993, 50). Augustine posits a second differentiation between two kinds of light. *Lumen* is the objective, inexhaustible, intelligible and divinely created radiance passing through and illuminating space. *Lux* is lumen's earthly, human reflection — our physiological experience of light and our capacity to receive it. Man, therefore, becomes a light lit by light (43), and the connection between the eye and free will begins to be established.

Belief in the equation "seeing = knowing" has a basis in classical Greek thought. Augustine's suggested relationship between the human eye and free will can be linked to his Platonic respect for a divinely inspired geometry whose reductive powers are made apparent via the eye acting as an agent for free will, as an intellectual mediator, *and* as a metaphor. Augustine's Platonism, however, also allows him to stress the primacy of geometry over perception. The eye is central to geometry's ascendancy; however, Augustine critically distinguishes between vision and sight. In his words: "Reason advanced to the province of the eyes. . . . It found . . . that nothing which the eyes beheld, could in any way be compared with what the mind discerned. These distinct and separate realities it also reduced to a branch of learning, and called it geometry."[6]

Into the Light

In contrast to Augustine, early Christian Neoplatonist mystics had conceived the finest access to truth as seeing *into* the light, as light was believed connected to the infinite and Heaven, and not of this earth. This effort at "direct perception" reflects the suspicion of *logos* and continues to enjoy support in our own time. When one *sees* an object, "it can appear without the fact of its appearance already involving intercourse. . . . [there is in vision a] clear separation between the theoretical function of information and the practical conduct, freely based on it" (Jonas 1982, 145). VR, often in the name of efficiency, indicates the ongoing search for prelinguistic communication practices.

The suspicion of *logos* directed these Neoplatonists to give themselves over through direct perception to be dazzled by *lumen* in as unmediated a fashion as possible. Ironically, however, given belief that one looks *into* the light, the critical distance implied between seer and godhead, between receiver and Sender, demands a conduit for mediation, no matter how much its presence is decried by such mystics who are the culture-denying precursors of VR's promoters who advocate "direct perception."

Medieval Neoplatonism continues the tradition of an earlier Neoplatonism that had reversed the original Greek positioning of the seer *in* light, and within the "wherein" of nature. The medieval seer looks *into* light in hope of entering its truth "therein," or "out there." Yet medieval light is a cobbling of this understanding to theories of light advanced by Cicero and Augustine. Medieval light is internalized to prevent "the worldly dark from fully penetrating and disempowering the subject" (Blumenberg 1993, 51). The cavelike monastic cell becomes the bulwark of culture and the recess of memory (Carruthers 1990, 40). Something like a memory trace of Platonic light is carried within, while the barbarous reality of a natural world from which the light has been withdrawn is sealed off from view.[7]

The medieval monastic memory could not resort to collapsing distinctions between nature and culture lest it perish in a "barbarous age." Where else to locate survival other than in interior retreat? Nicholas of Cusa (1401–1464), for example, reinterprets Plato's cave metaphor as the internal "ground" of the world[8] — a Platonic matrix of energies where human and divine minds intermingle, yet one that also suggests a relativity of perception. In a cave of one's own, truth is openly present. Inspired creativity is most easily found therein, and this intellectual activity, which preserves cultural memory as best it can, allows the truth seeker to emerge and "progress" in a forward direction (Jammer 1969, 39) along the opticized, instrumental path already foreseen by Roger Bacon.

Although the power of visual metaphor is diluted during the Middle Ages, Bacon's *Opus Majus,* written during the 1260s, petitions papal authority to redirect Christian inquiry in accord with a *vision*ary perspective (see also chapter 2). Bacon, in placing vision directly on an axis of truth, follows Augustine and elevates the status of geometry and melds it with embodied vision. *Opus Majus* reflects the thirteenth century's interest in optics and mathematics that followed the renewed influence of Neoplatonist thought, and its conception of space as infinite and open (Jammer 1969, 39).

Robert Grosseteste (c. 1168–1253), a mathematician and philosopher influenced by Neoplatonism, participates in the renewed interest in vision and optics that arises during the late-medieval period. Grosseteste holds light to be the first principle of the universe. Light can be transformed into other elements and gives intelligibility and motion to the universe (Reese 1980, 204). Grosseteste believed that the creation of the universe was the same as the self-diffusion of light, and that light *formed the basis of extension in space.* This led him to conclude that a true grasp of the universe was available only in the study of geometric optics (Jammer 1969, 38). This Renaissance interest in optics is also expressed in a revival of interest in Plato, and in the rediscovery of Euclidean geometry and Ptolemaic perspective. These, in turn, inform philosophical considerations of the world and have practical applications in mapmaking, perspective painting, and optical technologies such as Porta's camera obscura.

Victor Burgin's (1989) description of the Renaissance's synthesis of Euclidean geometry with the idea of a primary *perspective* suggests ways in which this synthesis fueled the development of a parallel connection between an absolute light on high and the slowly emerging inner light of subjectivity. Burgin sets forth differences between two of Euclid's works, his *Elements of Geometry,* which codifies a number of earlier theorems that conflict with one another, and his *Optics.* It is in *Optics* that the "cone of vision" is first theorized. In 1425 Brunelleschi theorized this cone to intersect with a plane surface, as part of devising his single-point perspective program. Although Euclidean geometry suggested an absolute and infinitely extensible 3-D space, the cone of vision helped to establish a somewhat contradictory belief that this infinite space had a center. Imported into single-point perspective technique, the cone suggested that the observer was at the center of space (15), as is the case with the panorama. Merging the cone of vision with perspective creates an instrument for subjective action, by which each observer is at the center, in possession of a light that with practice can be directed outward or inward, forward or backward. This inner light illuminates an individual vision that extends outward along the infinite coordinates of a geometric and mental grid conceptually stamped onto the Earth's surface. After Brunelleschi and the cone of vision, this inner light traces the sight lines of perspective across an infinitely extensible space over which the individual eye may imaginatively voyage as if on high, as if objective light itself. The inner world of VR creates an imaginary space for further extending this interior voyage, suggesting that the interior is infinite if not eternal—a hybrid proposal that collapses the dynamics of sight with geometry.

With human agency placed at the center of the dynamics of sight, the Platonic meaning of absolute light—more connected to vision and metaphor than sight and physiology—is inverted. The subject is also moving toward center stage and *into* the light. The unacknowledged articulation of geometric rules of optics and physiology of sight is given an overlay of self-consciousness. Plato had argued the moral necessity for ascension into the Ideal light so that humans might possibly attain full being — metaphorically, ascended truth seekers would achieve the widened array of ideals and forms presented to them within the refined, geometric space of ideal vision. The Enlightenment does not so much change Plato's imperative as reconfigure the spatial metaphors by which this human duty is given direction. As suggested in the discussion of the camera obscura in chapter 2, seekers must now labor to find the light within, and the resulting inward orientation helps explain the primacy of modern subjectivity (Taylor 1994, 29).

There are intimations in Cicero, more fully articulated by Augustine, that free will and the eye have a role to play in the production of light or the Good. With the eighteenth-century Enlightenment belief that humans are endowed with a moral sense came the understanding that this source of the Good also lies deep within each of us. Similar to how the early-modern subject within a camera obscura produces meaning within an interior recess, the self comes to be seen as harboring a separate luminary power from that residing "on high." This inner light is a metaphor for the Good, and the camera obscura and magic lantern are metaphors confirming different aspects of the belief that an interiorized light now shines from within. This difference can also be theorized by suggesting a parallel between the divine *lumen* and the camera obscura; the camera obscura is a technology of *lumen*. It truthfully reflects the objective world of exterior reality once wholly divinely given, though, for early-modern subjects, also culturally authored. In contrast, the magic lantern, and its world of shadows and illusions, is more a technology of *lux*, or *lumen*'s earthly, hence potentially more faulty (and sensual), reflection. As the interiorized subject increasingly positions herself or himself as the producer and judge of truth, however, the distinctions between *lumen* and *lux* become less hard-bounded.

Moreover, the early-modern viewer, whether having recourse to a camera obscura or magic lantern or both, does not yet have the wherewithal or cultural need to imagine that he or she might sublate his or her identity to the light as a condition for imaginative entry to an immaterial virtual world constituted in luminosity, pixelation, illusion, and information as data — a world positioned imaginatively as more truthful than the supposedly exhausted "real" of the natural world. Ironically, aspects of early-modern scientific thinking nonetheless tend to support such imaginative twentieth-century sublation. With Isaac Newton, the fundamental unity of matter and light is asserted (Koyré 1957, 207). Newton's unity can be read two ways. If light is matter, then VR's metaphysical edge is potentially muted. However, it is equally possible to conceptually relocate materiality to an optical "wherein." Since Max Planck in 1900 and Albert Einstein in 1905, if light is a form of wave motion *and* a fast-moving particle or "packet," then it is possible to envision an optical immersive technology such as VR as rendering communication seemingly concrete even as VR dematerializes the physicality of the world it represents into a transcendent and luminous "wherein."

If it is accepted that modern thought retains a variety of subtle Neoplatonic influences, this distance between the self and the true light (of God) one seeks then requires a conduit. The conduit metaphor of communications implies and demands uncontaminated passage of the message (from on high). Mediation across distance becomes the essence. When the metaphor of looking into the light is incorporated into virtual technologies, its earlier Neoplatonic moral function, which maintained and required purity of communication-at-a-distance from God to humanity, is updated and maintained by the assertion that technology is value-free.

With respect to Neoplatonic mysticism and dazzlement, no one is able to accustom oneself to the latter's absolute intensity, by which one is illuminated and blinded, has one's eyes fully open and resolutely shut. This mystical ambiguity was taken by early Neoplatonism to confirm God's illuminating and transcendent presence, which bypassed human communicatory and intellectual processes (Blumenberg 1993, 45). To be dazzled is to be flooded by the universal light of God — a state of "direct perception" achievable only by suspending the reflexivity and critical distance that normal cognition operating within a cultural milieu provides. Yet at the same time as this metaphysical directness-at-a-distance is being constituted as an axis for faith, Augustine also argues that one can open one's eyes in the dark or close them to the light, turning one's gaze inward. This free will in part depends on the light that increasingly comes to be seen as "shining within" as a reflection of God and therefore "above" or "before" culture. Sight-dependent subjectivity had been absent in classical thought, which, in its various metaphors of vision, had not accorded this degree of primacy to the eye. The interior self coming into being in Enlightenment thought is fertilized by philosophy's elevation of the eye's power, which is made to operate within the opening starting to develop between nature and culture in post-Hellenistic philosophy.

As if to anticipate individuated experiences within immersive virtual environments, Ciceronean and Neoplatonist direct perception attained by contemplation of pure light is an act of splendid isolation, and perhaps only conceivable within a sphere of dazzling *luxury*[9] and culture, one that exerts a geographic or material influence on philosophical formation, the sociopolitical effects of which are often ignored. However, once one begins to communicate not only with God but with other people as well, the relationship between purity and the conduit metaphor of communication must evolve. If a (vertical) pure conduit was needed to transmit God's word in as uncorrupted a fashion as possible, thereby eliminating "noise" from the heavenly transmission, when the conduit becomes "horizontal" — running between mundane, imperfect places — then the conduit's purity implicitly remains available to purify the message being transmitted to "imperfect receptors." The reference is elevated above that to which it refers or from whom or where it was sent. The conduit or the technology, then, not only is potentially value-free but is further privileged as morally superior to the message, sender, and receiver. To communicate through a medium, therefore, is to have a sense that one's message might be touched by God. Enter metaphysics, and the more light-dependent the technology, the more metaphysical the uses to which it might be put in seeking "truth." As if anticipating, for example, the pseudoscientific New Age celebration of channeling's "ability to resolve the technical problems of communication" (Ross 1991, 37), the purity of Neoplatonic light-as-conduit spiritualizes information and the means of enlightenment and communication (see Davis 1993, 612).

In and of the Light

I have made links to VR throughout the foregoing history of light. I am now in a position to argue further continuities between mutable metaphors of light and VR and VEs. In gestalt terms, immersive virtual environments marry the "modern ground" of an articulating Cartesian grid to a "field" of the polyvalent "identity formations" they situate. They are defined by light, the informing essence of sight. Given the spatial ambiguity that attends the Platonic "localizing" of truth-as-light in transcendency, *where* such a transcendent locality might be found would seem destined to remain a perpetual mystery. However, localizing truth in transcendency via light implies that movement and, by extension, communication and its technologies become ironic sites of truth in and of themselves. In the West, such movement is often related to the emanating power of light. Truth becomes linked to movement and the pure, immaterial, and Ideal "space" of communications. VR confirms the radical disconnection from real places with the modern practice of looking inside oneself for the light of truth. Real places are made to seem beside the point when truth is "localized" to fiber-optically dependent transcendency, motion, luminosity, and "flow."

Given the high status accorded interior subjectivity and self-identity, virtual environments may be seen as a relocation of the absolute light from on high to a place more convivial to this inward orientation—a kind of super-nature and peep show rolled into one. Yet virtual environments are also a pure, "interiorized space of culture" not unlike "where" we are now expected to find our own guiding lights. There is a simple progression here. Light is first on high, in the sky like God and the sun. Later, as nature materially recedes from cultural purviews, light relocates *to* the sphere of culture and then even inside ourselves *as if* we were gods. As technology is strengthened, we are able to relocate this light to within optical technologies that confirm the "naturalness" of the inner light of individuated subjectivity.

The Western "condition" has accustomed itself to the naturalness of establishing a connection to the world by viewing it; "we do not so much look at the world as look *out at it,* from behind the self" (Cavell 1971, 102). With VR, users must first of all "approach" the technology, a spatial move familiar to seekers entering the light *or* looking into it. In Neoplatonic fashion, as users don the HMD, they look into a virtual world composed of light. However, by then relocating a part of these individuals' sense of self to an icon located both in and of the light, the technology collapses the Neoplatonic distance between light and self. This collapse is already under way with the stereoscope; however, by positioning the seer in and of the light—as both illuminated and wherein—VR goes beyond the stereoscope to suggest a transcendent doubling and collapse: both it and that part of the seer's iconized self "within" the technology now form a natural place. In the words of VR researchers Richard Held and Nathaniel Durlach: "Taking liberties with Shakespeare, we might say that 'all the world's a display and all the individuals in it are operators in and on the display'" (1991, 232). For Held and Durlach, it would seem as though users have become one with the program. This suggests the potential that spatial distinctions between a user's self, or sense of self, and the world around her or him, could become volatile and unstable. Implicitly, however, Held and Durlach anticipate no problematic out-

comes issuing from VR collapsing the distinction between humans seeing the world around us and it showing itself to us. In any event, VR organizes a different binary in which users interact with images they can, in
some programs, alter or design, but only according to the preconditions
designed into the technology. VR monitors users' body motility in part
to reconfigure the images it presents. These images, however, are at least
partially authored by the technology's designers and are subsequently
translated into the code on which VR relies.

Virtual environments, therefore, are also a pure interiorized space of
culture, the virtual "poststage" stage "where" we are now expected to find
and also be our own guiding lights. Recalling the contributions of the
magic lantern toward thinking about virtual worlds, it is worthwhile to
consider the emphasis on phantasm/fantasy accorded the magic lantern
in contrast to the aura of science and truth that bathes the camera obscura. A nineteenth-century magic lantern experience took place in a
cellar or darkened room and relied wholly on artificial light. The experience subverts the meaning of Plato's cave — a metaphor promoting separation of the faculty of sight from true knowledge. It is also worth considering that too much artificial light or *lux* deflects the quest for truth
through the use of light into a pursuit of fantasy, today more commonly
subsumed under the rarely questioned rubric of pleasure.

The continued quest for technological progress is rooted partly in a
belief in human perfectibility — that the light or moral core of goodness
is within us, if we work to find it. Optical technologies — from the camera obscura, to the magic lantern, to VR — along with the metaphors by
which these technologies are discursively and strategically positioned
within culture, are applied to making this exalted task of getting in
touch with the light less onerous, and this alone becomes adequate moral
justification for the current focus on virtual transcendence machines.
Despite their differences of scale, both tools and technology extend our
grasp. Earlier tools — such as the metaphorics, geometries, and other "visible instruments" that, for example, Roger Bacon believed would more
fully reveal "the form of our truth" and "the spiritual and literal meaning
of Scripture" — were intended to access God. For later Enlightenment
theorists postulating a light within, the fixed source of Absolute light
above was not fully extinguished. Although for many today the God behind this light is missing, lost, or "canceled out," optical technologies
may be seen to offer a labor-saving substitute: to allow individuals to
communicate with one another as participants within an ideal sphere

of continuously circulating communication. The wish to achieve such a state is what drives Kevin Kelly's (1994) synthetic vision of telematics, metaphysics, and politics. Kelly is editor of *Wired*, the successful mass-circulation magazine promoting the telematic reality of a coming wired world. He proposes we join together as "dumb terminals" in an ecstatic unity via a rhizomelike cybernetic net to achieve a state he identifies as "hive mind." Hive mind is a techno-humanist version of the ancient notion of "world soul." For Plato, this was the principle of animation in all things—not unlike the power of the immaterial "wherein" of light. For the Egypto-Roman Neoplatonist philosopher Plotinus, world soul was an emanation of God, and the physical world was God's body. For Kelly, hive mind is the collective buzz of networks in which all bodies have become informational. If there is neither a natural location for the cosmic soul nor an otherworldly God available to Kelly today, there is the substitute possibility of fantasizing the fiber-optic and light-dependent technologies of the Net and VR's spatial display as the immaterial, utopian embodiments of information as deity—a transcendent nowhere location "where truth has gone."

In a sense, virtual technologies collapse older distinctions between the sacred and profane—between the quest for truth and the desire for fantasy. The sublime netherworld of information permits the optical illusion that human bodies might merge with computers and the light "within." To quote the boy wonder computer whiz Bryce Lynch, from the TV series *Max Headroom*, "You're looking at the future, Mr. Grosman—people translated as data."[10] To achieve such an incantatory state would be to ward off all the real-time viruses and other plagues of the flesh that bedevil those who would gladly take leave of their "impure" or profane earthly form and their rootedness in the here and the now. This impurity, for those like Kelly seemingly a curse, is constituted in a failure of the senses, belief in which can be traced back at least as far as Descartes. "I think, therefore I am" is the opposite of the "dumb terminal"—a body or *automata* that communicates in a defective manner and therefore needs prosthetic devices to extend it toward enlightenment.[11] The purity of the conduit that Neoplatonic light demands to transmit its divine message from Sender to receivers is conflated in hive mind into the network's almost sacred ability to resolve the "problems" of embodied, sensual communications. Held and Durlach's functionalist reduction of human experience to operators in and on the display shares the logic of hive mind and reflects the ongoing wish that somehow communication technologies might both illuminate and stand in for embodiment and the onto-logical ground upon which (we think) we stand. Virtual Reality: to enter magical empiricism's world of *as if.*

Technics, Theoretics

VEs depend on language and code. As part of considering how theory and technology play off each other to suggest the collapse of real places into language practices, I will examine Mikhail Bakhtin's theory of the carnivalesque marketplace and the homology he suggests between mar-kets and the modern print novel. The theory of carnival is relevant to my concerns for two reasons: first, it relies on a metaphysics of language in asserting its politics of individual freedom; second, its construction as an academic argument and its more recent exhumation by certain strands of academic postmodernism parallel the development of VR, the post-modern technology and the claims advanced for its liberatory potential.

As representations of space, VEs act as built metaphors of light and vision. Within a pluralist society, there will always be assessments made by some that metaphors are being misused by others. Such "misuse" — the accusation of which always has political overtones — seems inevitable. Out of respect for civil public discourse, cavalier use of metaphor ought to be guarded against as best one can. Although I am highly sympathetic to his aims, Bakhtin's "carnival" exemplifies an ill-considered use of met-aphor. Bakhtin's theory of carnival, developed in *Rabelais and His World* (1984), "translates" real marketplaces into textual form and does so in a way that collapses material or formal differences between real places and textual representations of them. The "carnivalesque" medieval market-place brings together poor and rich, peasant and lord. The collective belly laugh of the crowd — the public performance of gross bodily functions, oral utterances, and speech practices — levels social stratification, thereby offering the oppressed a temporary taste of political agency. However, according to Bakhtin, carnivals were not only places where goods were traded and provincials might obtain a glimpse of a mobile cosmopoli-tanism. They were also subtly sanctioned spatio-temporal means of vent-ing steam so that whatever hegemonic state of feudal affairs existed, how-ever gruesome, might resume operations after its brief abeyance. Bakhtin links the embodied carnivalesque belly laugh, the "low life" of an ironic and illiterate peasantry, to the diegetic possibilities contained in the literate form of the modern novel, with all its attendant cultural capital. Bakhtin's exhumed medieval carnival can only be transubstantiated into the novel's abstract textual representations when voided of embodied speech practices. Ironically, these earthy practices that so impress Bakhtin are seen by him as relocatable to the novel form. He thereby asserts a correspondence between carnivals and novels and the agencies they situate.

By idealizing medieval carnival, Bakhtin aims to recapture a Rabelaisian moment of hope, one that he wants readers to remember, even as he understands this moment is no longer an embodied part of humanity. Perhaps because he grasps the modern impossibility of Pantagruel's Utopia, Bakhtin fuses the embodied carnivalesque — the belly laugh and the fart in the face of noblesse oblige — with the dialogic and formal possibilities of print. However, to do so, he employs *paralogism* — an excessive use of metaphor that skirts issues of form and scale and suggests the interchangeability not only of historical epochs but also of physical embodied realities and their representations.

The medieval carnival, *when voided of embodied speech as spoken, oral language* — of those ritual aspects of communication noted by Carey (1975) and discussed in chapter 2 — is made available for conceptual relocation to the print-dependent novel, whose abstract textuality and physical inertness in the reader's hands may allow for critical self-reflexivity. The public square of carnival — a place that brings together the continuity of local identity with the possibility of change as represented in the wares, strange customs, and stories of traveling salespeople — seems even more readily conceptually relocatable to virtual public spheres precisely because of their greater iconographic and linguistic naturalness than printed text.

For purposes of the present discussion, Frances Barker's 1984 study of the effects of the Restoration on the social and political imaginary of 1660s England offers a useful counterpoint for considering the implications of Bakhtin's theory. Barker finds that during this period, human bodies ceased to be a public spectacle and were privatized in novel ways. The subject largely abandoned public performance, coming to terms, instead, with representing "itself" in text made possible by print technology. The modern silent watching of theater as spectacle exemplifies this withdrawal of the subject from "carnivalesque" performance. At the same time, the bourgeois subject found itself opened to new forms of manipulation by virtue of its spatial and mental isolation. Traces of this isolation are found in the extreme forms of privacy that gestate behind locked doors. The Cartesianized self retreats from public display and increasingly relocates its expression to texts. In a way that Bakhtin never considers, Barker recognizes the synergy between changing forms of communications and changes in social relations: from the performative "song and dance" of the communicating public body, to its representational and imaginary reduction to a commodifiable media form.

Bakhtin privileges voice as a political act. He wants to demonstrate the commonality between embodied speakers and listeners. Yet his theory also suggests a threefold and slippery interchangeability or formal equality among (1) utterance and vocal sounds, (2) the words we speak aloud to one another as part of our presentations of self, and (3) these words subsequently codified within communication technologies such as print and ITs. However, there is a difference between physically performing in carnival's boisterous world of mirth and reading about it silently in a room of one's own hundreds of years later. This spatio-temporal difference matters sensually, formally, economically, and politically.

Theories according primacy to textual representations, such as Bakhtin's and more recent postmodern and poststructuralist notions, are often deployed to assert that (1) there is no (knowable) world beyond the text, (2) belief that representations relate to the real world constitutes a "category mistake," and (3) identity is a futile concept, as we are consigned to repeat performances by which we momentarily confirm who we are. Such theories resonate with and support the dominance of telematics and IT. In a variety of ways, as technologies and practices, information technologies confirm not only that "all the world's a text" but also the cogency of these academic theories, which are themselves partially extended, as in the case of Baudrillard's understanding of the simulacra, from information theory. Such theories, along with VEs, not only ignore distinctions between the power of pictures and that of more abstract textual representations but also are premised on communicating agents extending *away* from bodiliness. Instead, a disincorporating subjectivity is actively directed toward metaphoric "spaces" and representational language practices, within which a kind of purified carnivalesque liberatory potential might seem to reign free.

Such inattention to the meaning of form skirts an abandonment of ethics. In Elizabeth Grosz's 1994 project to advance or reintroduce a more holistic understanding of body and mind and thereby reinvigorate contemporary feminist theory, she writes that Descartes's principal achievement was the "exclusion of the soul from nature" — the fabrication of a binary that allowed "evacuation of consciousness from the world" (1994, 6). Now, Walter Ong (1991) has focused on the relationship between orality and technology, paying close attention to the changing medieval European speech practices that would have informed Rabelais's world, which was in an uneven state of transition from an oral to a print culture. Concerned with the impact of print on the development of an interiorized modern consciousness, Ong writes:

> By removing words from the world of sound where they had first had their origin in *active human interchange* and relegating them definitely to *visual surface*, and by otherwise exploiting visual space for the management of knowledge, print encouraged human beings to think of their own interior consciousness . . . as more and more thing-like. . . . Print encouraged the mind to sense that its possessions were held in some sort of *inert mental space*. (1991, 131–32; emphasis added)

Reading Ong against Grosz, "inert mental space" must first be readied to receive consciousness as it is evacuated from "active human interchange" with the world and nature. Over time, interiorized consciousness (secured "behind" Elias's wall and other less metaphoric devices such as, for example, gated communities and their reliance on communication and surveillance technologies) seeks to extend itself "outwards." Desirability of products such as VEs, which allow disembodied communication that seems to penetrate across the wall, will grow. Such products substitute for the "positionality" required of an earlier fleshy collectivity. They act as post-Leibnizian windows admitting vision and light onto a monadlike consciousness, so that each self, alienated from nature and collectivity, might communicate the relative fact of its lonely and conceptual existence to other equally disaffected selves. This disaffection operates in tandem with the very move of consciousness toward inert mental space and the loss of skills implied by the ceding to mediation of embodied human interchanges that can facilitate and advance recognition of differences between people. Such skills depend on the myriad face-to-face embodied communication and speech practices between strangers in public, partly required to take place within an increasingly fugitive "public square."

The disappearance of the polis as a site for political performance in favor of the creation through technology of a mediated public sphere that corresponds to the "public square" it appears to render antique is a victory of applied technology over politics. A VE brings together the power of applied science and the meaning of language to construct a "picture language machine," suggesting that reality might be entirely a discursive formation. Equating public performances with their representations — without giving sufficient attention to the ownership of access and the commodity forms that users must adopt — is a naturalizing use of metaphor; it obscures a process whereby aspects of users' independent agency are given over to forces controlling the technology.

As communications technologies, VEs drain the meaning of form and distance in a manner similar to that performed by Bakhtin's theory. Like the theorized voice of the novel cum carnival, VEs suggest the technical feasibility and hence cultural acceptability of conceptually relocating the disincorporating self to a reality constructed from a visibilized language set apart from the human body — as if the subject might somehow physically "move into" language itself. Philosophically speaking, an idealist belief in the possibility of a world constructed entirely from technologies based on language would seem to promise a pure, if unanticipated, victory for social constructionism. This would be a Pyrrhic victory, however, one in which "culture in effect takes on all the immutable, fixed characteristics attributed to the natural order" (Grosz 1994, 21), and one in which language itself would become a pure spatial technology. But VEs achieve something more in their metaphysical aesthetic pursuit of a "perfect copy" of reality. They collapse distinctions between figure and ground, between bodies and space, and thereby also the possibility that body icons in cyberspace might really stand out against, or be defined in terms of, cyberspace.

One might argue, therefore, that in VEs two alternate scenarios are possible: First, the ground upon which bodies "stand" is collapsed into the figure or subjectivity of the viewer, a dynamic suggesting that all of the natural world is a cultural product, including the "self." Second, one might reverse the argument to say that subjectivity is denied in being made to merge with the ground — a kind of rediscovery of the wider world reformed as information.[12]

Yet both scenarios are entirely representational. They suggest the much ballyhooed cyborg, which in cyberspace takes the form of a quasi-metaphysical and electro-flesh merger between (interiorized) users and an essentialist or absolute conception of (exteriorized) space. This merger operates within a material technology reflecting the technical elite's wish to discard its flesh in an act of body denial disguised behind depoliticizing metaphors of transcendence and liberation. This denial fuses with a sense of feeling overburdened by the duty to create self-meaning by a self-consciousness that confuses itself with "mind," along the way confirming to itself the dumb animal status of its own body materiality. Both scenarios also assume that a direct correspondence exists between reality and its models. Applied to VEs, both scenarios assume that all reality, our bodies included, can be reduced to representation — in this case a two-dimensional one attempting to pass for 3-D. Regrettably, improvisation — creative response to contingency and externality, and using "what is at hand" to do so — is difficult if one is trapped "within" the predeterminations of a model or theory.

Ken Hirschkop, in answer to his own question as to why Bakhtin declines to revel in the difference between the novel and everyday conversation, suggests that in Bakhtin's world, "the brute facts of modern life make real dialogue unworkable" (1991, 110). Faced with the withering of free speech, Bakhtin saw the novel as the best of a limited array of choices for preserving independent expression of thought during the Stalinist imperial terror (see Anchor 1985, 238). In an authoritarian state, the printed page disseminates oppositional information more surreptitiously than embodied free speech. Yet for Hirschkop, and all those more fortunate to operate within a less dangerous politico-intellectual climate than Bakhtin, an attention to the different locations, strengths and forms of "genres which cite and represent" [and] "the public square" (Hirschkop 1991, 110) is essential to any argument that claims to be about theory: even if, I would add (or perhaps especially because), the opportunity is available to identify links (including metaphoric ones) between theories such as Bakhtin's and the political philosophies they support.

Robert Anchor is correct to draw attention to the difficulties facing Bakhtin as a dissident intellectual voice. However, the dissidence in *Rabelais and His World* is more than only the critique of Stalinist repression achieved by describing the freer exchanges in medieval carnivals than in 1930s Leningrad and Moscow. Bakhtin excels in writing about the hybrid and "grotesque" admixtures of high and low, politesse and buffoonery, and he is unsparing in his criticism of modern writers who make over the grotesque into the "other" and thereby completely expunge it from (their own) sanitized, rational consciousness. The grotesque hybrids Bakhtin finds resident at market carnivals highlight the socially constructed distinctions established, maintained, and finally "naturalized" between high and low culture (Stallybrass and White 1986, 39).

Nevertheless, Bakhtin's grotesques are situated in a context that now is *past*. For him, the "real world" — Stalin's Soviet Union — is a Platonic illusion of flickering shadows. Only in the past, or "in" the modern novel, can the true and ideal light operating as an act of communication be found. In any event, for Bakhtin the two may as well be the same, and this he achieves in his representation of carnival, where every social stratification is leveled in an ideal utopian *moment* frozen in time and space. The past, like nature, is made unavailable to human beings, and hence it must be represented. Peter Stallybrass and Allon White note Bakhtin's "collapse of the fair into the literary text and vice versa" (1986, 60). Bakhtin's carnival mystifies the social relations of rape, anti-Semitism, and commercial exploitation present in the original as readily as it "sheds light" on a utopian, revisionist, hope-inducing *image* believed worth "holding in view."

Stallybrass and White write of marketplaces as common places, as the "epitome of local identity" yet the "unsettling of that identity by the trade and traffic of goods from elsewhere" (1986, 27). They criticize Bakhtin for removing markets from the effects of territoriality and note that real places are not the "utopian no-place of collective hopes and desire" or the "pure outside" (28) that Bakhtin's text-dependent marketplace has become. I would note that real places, actual carnival markets included, hold together the conceptually discrete categories of nature, meaning, and social relations. They contain within their leaky boundaries the dynamics of continuity and change. Place is where the "grotesque" can never entirely be alienated from *mundane* experience, or made into an "other" that fits no category and is consequently shunned.

Bakhtin argues that "languages are philosophies — not abstract but concrete, social philosophies" (1984, 471) whose structures communicate the latent liberatory messages within them (Hirschkop 1991, 105). The inherent structuralism here — that dialogical interaction is built into the very structure of language (104) — is similar to James J. Gibson's assertion of the perceptual invariants and their directly perceived wealth of information. For Bakhtin, language becomes a political act in itself, or if not this, then the site of such an act. Dialogism refers to the conversation and dialog always already instantiated by language. Paralleling Dreyfus's comment that the West makes technologies out of its philosophies, arguments proclaiming the liberatory potential of interactive immersive VEs bear a striking similarity to Bakhtin's theory of language. Both the labors of a novel's author and the efforts of backroom techno-wizards collectively writing visible virtual worlds for the contemporary electronic marketplace issue from a "structure of feeling" (Williams 1960) and necessarily carry within them an array of cultural biases. The concepts of dialogism and interactivity can mystify the fact that readers and users must also "dialog" or "interact" with what is placed before them. While a *range* or *array* of discussion and choice may be intended by the writer or inventor, it is somewhat determined in advance, though rejection or reinterpretation of the original intention always remains possible. However, the military VE described in the introduction, which permits users to assume different tactical positions within the closed world of a virtual battle zone, does not alter the warlike context within which the various positions are arrayed. Freedom to move about the space of a virtual battle zone or assume different identities "therein" is not the same as changing the program, and the surveillance mechanisms written into it, to one that, for example, simulates a dialogue of peace.

Assertions about the nature of dialogism and interactivity are suggestions that to "consume" — a novel or a simulated experience of reality — is not really much different than to "produce." This avoids addressing the nature of work as it minimizes distinctions between "making meaning" from reading and "making meaning" through writing. Although consumption and production of texts form a dialectic of communication, they do so in subtle ways and through different contexts that are inadequately theorized by the notion of dialogism alone. When assertions about the nature of dialogism are related to how the imagination is stimulated, the claims are less dubious than when they are translated into, or read back onto, material forms. Attention to the substance differences between real places, printed pages, and pixelated spatial displays suggests the fallacy of believing the original assertion can survive the translation across the form. The bridge of metaphor is strong enough for the imagination but not always for the body to cross.

In Chapter 4, I criticized the belief that VEs based on James J. Gibson's direct perception theories might offer a full correspondence to material reality. I noted the distinction between the language of facts and the language of states (Dretske 1969), and Burgin's (1989) injunction against mistaking observable behavior for psychic truth. I also noted that to rely on lawlike "correspondences" alone forgets the importance inherent in differences between the substance of things (Sack 1980). About ITs, Lyotard (1984, 4) writes, "the nature of knowledge cannot survive unchanged.... it can fit into new channels... only if learning is translatable into quantities of information... anything that in the constituted body of knowledge that is not translatable in this way will be abandoned." Lyotard seems to suggest here that knowledge and epistemologies — akin to Dreyfus's philosophies — get made into technologies or "new channels" if the knowledge or philosophy is already in some way amenable to the "translation" or making over. Mark Poster (1990, 70) dismisses assertions "that nothing significant is lost in the process of digital encoding, storage, retrieval, transmission and reproduction." I do not take Lyotard's and

Poster's observations as antithetical to Dreyfus's, for the nature of form and the differences between forms *require* a change when such transitions or translations are made, even though the original intent or *desire* may have been to minimize the unavoidable changes in form along the way from philosophy to built technology.

> It is not the case that discursive material is transmitted intact between existing, fully-formed discursive spaces which act as donors or hosts.... Sites and domains of discourse, like ... [Bakhtin's] study of the marketplace ... [emerge] out of an historical complex of competing domains and languages each carrying different values and kinds of power. (Stallybrass and White 1986, 60–61)

The laughter of carnival, as a means of experiencing our common humanity through a momentary collective transcendence, is *theoretically* transposable as data to a VE engineered to represent a spatially isolated and individualist view of social relations. The imaginative possibilities contained in the carnival may not be entirely absent in a parallel virtual world. A virtual world might *seem* to offer a sphere of freedom and equality subject only to the programmed laws of its own internal logic. Such laws, however, would prescribe language operating as a law unto itself, a subject examined in the following section.

An Architecture of Language

Form matters, both in this world, and in virtual worlds on the other side of the interface. In this section, I argue that the form of spoken language allows it to be thought of — and to act — as both a part of our "inner" subjectivity and embodiment and that which allows us to extend ourselves into the lived world around us. Language also allows us to blur, or hold distinct, differences between "an idea of the body" and actual bodies. Held apart, ideas and bodies can be made to support distinguishing between culture and materiality. If connected by language use, however, the relationality between idea and matter becomes more imaginable. In other words, language, like bodies, straddles a line between the modern distinctions erected between nature and culture and can be thought of as one of the earliest hybrids or even a prototechnology. As a representation of language, writing straddles no such line. Writing passes, thinglike and discrete, into a purely cultural realm. VEs, dependent on a base of written codes, are a form of writing practice that trades on the social bonds that language establishes and extends. Their cultural point of purchase partially depends on the necessarily imprecise and fluid distinctions and collapses between writing and language, idea and materiality, and culture and nature. As metaphors, VEs are positioned to work both sides of what they seem to conjoin.

"Bakhtin recognizes the duality of every sign in art, where *all content is formal and every form exists because of its content*" (Pomorska 1984, viii; emphasis added). However, especially with respect to the italicized sections of Krystyna Pomorska's assertion, the opposite seems the case. As she implies (and as I have stated in linking Lyotard's argument about knowledge technologies to Dreyfus's assertion that the West builds its philosophies), the interrelationality of form and content is precisely why shifts in use from one cultural technology to another matter. The relocatability of Bakhtin's medieval marketplace to the modern novel is a strategic move not so different from Jaron Lanier's holding forth the image of actually building a house in a VE as an instance of the direct creation of reality. It is true that Bakhtin attaches great moral and political weight to the structure, or form, of language within the act of communication. Lanier does too. Yet both believe that the exchange of messages is an end in itself. Home alone, reading about carnival is rendered equivalent to having participated in the parry and thrust of a now dormant form of public square, as is the imaginary making of a house within a VE claimed to constitute an actual experience of house building.

Conjuring carnival from print, or a house directly from an imaginative blueprint without the intermediate necessity of physical labor and materials, depends partly on a metaphysical blurring of the physical experiencing of one's body with an idea of "the body," and an inattention to meaningful differences between forms. A form's limits both constrain *and* enable its possibilities. With an ever greater emphasis on communication via representation alone, and the new forms of technology that advance this cultural direction, the original carnivalesque "laughter" Bakhtin wishes to transform into a metaphor of hope and political resistance is increasingly detached from its referent. Carnival's sound is transformed into sets of linguistic symbols "outside" the experiences of those described; what once was present to be heard, felt, and spoken is cloaked in the pattern of visible representations made exterior and "thinglike" to the self.

The novel occupies an intermediate position between spoken language and the iconographic world within virtual technologies. The latter rest on digital computation. Digital computation confirms the dichotomy "between perception and semiosis as two aspects of mind, and [it] comes down firmly on the side of semiosis. . . . Like all writing systems, the com-

puter must work through signs in order to represent, classify, and operate on perceived experience" (Bolter 1991, 224). Gerard Raulet finds the
computer is the "first machine that 'works' in language," and that [infor-
mation] "technologies directly possess the social bond." They treat language as capable of expressing every thing, including the ineffable, as "a
totally transparent utopia"(1991, 45). In possessing the social bond, ITs
may possess what is public between us.[13] Raulet echoes Bakhtin's expectation of language. It remains the expectation of VE engineers that a sufficient quantity of sophisticated hardware and software languages will
fully simulate representations of spatial realities along with our selves.
Raulet's linking of language to externalized "things" is notable, for implicitly he is writing about language not as a speech act but as something
independent of nature and society alike. Writing is a technology. "Print
suggests that words are things far more than writing ever did" (Ong 1991,
118). It furthers an already underway technical exteriorization of speech.
Now, VEs are cybernetic environments. In 1950, Norbert Weiner, credited with coining the term "cybernetics," theorized how humans might
relate to thinking machines.

> Language is not exclusively an attribute of living beings but one which
> they may share to a certain degree with the machines man has con-
> structed. . . . in constructing machines, it is often very important for us to
> extend to them certain human attributes. . . . If the reader wishes to con-
> ceive this as a metaphoric extension of our human personalities, he is
> welcome to do so; but he should be cautioned that the new machines will
> not stop working as soon as we have stopped giving them human support.
> (Weiner 1989, 75–77)

Virtual technologies make language visible in powerful ways. Cyberspace is the "becoming-visible of writing as writing . . . the perfect 'fit'
between technologies of writing and the body-machine complex" (Seltzer
1992, 108). To translate something of the meaning of embodied voice to
a VE has meant reformulating this voice as an exterior thing the eye can
behold.

All forms of writing are spatial. Each technology, however, uses space
differently. Earlier writing took place on the surface of a continuous
scroll; within the book format, print takes place on the separate and
bounded spaces of pages that are miniature territories or grids. Electronic
writing on computer screens again changes writing's spatial display. The
computer-language matrix — whether the symbols materializing on the
screen of a word processor as a semifinal product, or the underlying code

that drives the software—is yet more spatial than texts because of its further removal from the spoken voice (Bolter 1991, 161), and this matrix promotes a renewed interest in the "long discredited art of writing with pictures" (46).

Iconic self-representations in an immersive VE seem predicated on language expressed as thinglike digitized codes, algorithms, and software—autonomous from human subjects. But even before this, the inner properties of objects have been discounted. If this were not the case, how would it have been morally conceivable to have imagined that the quality of any object could be "duplicated" by its representation? In a VE, language is not only an expression of the body; it has become the principal formal content of an idea of the human body expressed as data image, as a specific form of information rendered as digital patterns.[14]

I am not suggesting that humanity might somehow exist *apart* from language, but that language is *a part* of the larger whole of our embodied humanity and any consciousness that attends to this. Language's oral and textual forms extend the human subject to engage the world beyond her or his body. A central demand placed on modern subjectivity requires maintaining an awareness of language's and communication's roles in making us human while at the same time requiring the subject to interpolate this awareness with a consideration that representation proceeds only on the basis of concrete bodies and other material objects, the simulacra notwithstanding. Performing this interpolation can be an ambiguous balancing act. I am suggesting that the value of this performance is in danger of being forgotten.

Language is not only a discrete, concrete thing, though certain formal *representations* of it can make it seem so. Neither is it ephemeral. Language can be thought of as an "embodied prototechnology," both confirming us to ourselves existentially at the level of embodied voice and extending us to engage with the lived world through its symbolic affect. Language precedes the sharp modern distinction between nature and culture. Extending Latour's argument about the modern hybrid, language, in its expression as oral speech, can be thought of as one of the original hybrids between nature and culture (so, too, is the human body). Language is a partial representation and extension of thought that nonetheless retains embodied form in voice—and this is the source of its synthetic power. To reduce language to a structuring mechanism of space inadvertently suggests that concrete reality is only a language construct. Yet if this really were the case, we would already understand spatial relations implicitly, as they would be entirely contained within language and the sphere of human meaning. This is not entirely the case, however much we might wish it to be so by asserting the suzerainty of possibly valid "generative grammars" or "perceptual invariants" that nonetheless are only part of "the picture." Such assertions have the effect of suggesting that space is only a conduit metaphor, and they further the move toward virtual technologies such as VEs, which are predicated on moving spatial relations into discursive formations modeled as conduits and containers.

Moreover, if we believe that telematics might adequately represent us in information space, we accord a "monopoly of knowledge" to representation. Raulet (1991) goes further and suggests that because "all becomes allegory" within delocalizing representational technologies, all therefore becomes relative. Beside Raulet's dire, potentially totalizing observation, I would still note Bolter's (1991, 235) finding that in a pluralistic society computation offers the only kind of unity now possible—unity at the "operational level."

Umberto Eco (1983) asserts that the Platonic space between the Ideal image and Real things has "collapsed." Although the cultural world is rife with signage, "there is still a great difference between signs viewed as a world and signs viewed as text" (Connor 1993, 78). James Connor theorizes a point similar to one I make in these pages in claiming that VEs are built metaphors when he asserts that VR "expands the text to a world." He argues that VR enforces "the ancient immediacy of self standing before an object, calling it real" (78). He is less concerned, however, with the technology's ability to mask its own representational nature and thereby conflate what it is to see with how this is represented. Users might well experience the immediacy of standing before a virtual object and calling it real. However, they do so only by ignoring their increasing spatial and experiential proximity with the technology, which becomes less and less an external thing-in-itself. If one looks at it in this way, one might even say that the hybrid or cyborg subjective technology that results is a building of the Kantian phenomena. Like Derek de Kerckhove's description of television-as-hybrid, contemporary identity comes to be "neither real nor fabricated, but both at once, neither public nor private, but both at once, neither inside me nor outside, but both at once" (1991, 268).

The contemporary paradigm of Gibson's perceptual invariants, and what this arrogates to the eye, would appear to have deflected the attention of some VR theorists and engineers from adequate consideration of the relationship between spatial displays and the thinglike language base underpinning the illuminated iconography the displays manifest. Benedikt's (1992b, 125) earlier noted assertion — made in the context of theorizing VEs but drawn from a consideration of the real world — that space and time together make up the most fundamental layer of reality does not address what constitutes the substance, however immaterial, of spatial representations within a VE. "Before" or "beneath" virtual space there is a "layer" of language that underlies the virtual world. Now, conceptually, space is conditioned by language, which in turn inflects our experience of geographic reality. Part of how we conceive of space depends on prepositions in language — the *in, at, to, of, into, out of, between, up, down, inner, outer, over, under, dans, en, dentro,* and *detrás* of speech. In other words, specific languages may prepositionally condition our experience of space in specific and arbitrary conventions.[15] As a theorist of cyberspace, Benedikt operates uncritically within an intellectual framework that accepts language as a law unto itself. He uses language to promote his understanding of VEs as a stand-in for material reality. He is saying that a "conception" forms the fundamental layer of reality. Whereas this is definitely true within a VE, it is arguably false in the real world, even if a variety of philosophies hold that there is no ultimate proof or disproof of this assertion.

6. Identity, Embodiment, and Place—
VR as Postmodern Technology

> In the absence of geography since we've explored the world, now there's
> the construction of new geographies through the computer, or through
> simulation, or through digitalization, or through replacement of the body.
> **Narrative voice-over from the film *Synthetic Pleasures* (1996)**

Immersive virtual technology seems to offer more real sensation than
older visual technologies for at least two reasons. First, it radically shrinks,
if not eliminates, the actual distance between the user's eyes and the HMD
screen to less than an inch. One's head feels thrust into the perceptual
field of vision. The second reason involves the technology's ability to facilitate the adoption, trying on, or acting out of multiple aspects of the self.
VR offers conceptual access to a space perhaps best appreciated by people
manifesting multiple personalities, and who, by their interest in VR, are
responding to cultural demands that fracture identities previously held
to be more unified (Stone 1992b). VR can be seen to support the fragmentation of identity and render proliferating individual subidentities and
their experiences into commodity form. A VE also provides a space of
performance, a multipurpose theater-in-the-round for the many components of the self.

In VEs, a quasi merger of embodied perception and externally transmitted conception happens at the level of sensation. The appeal of this
electronically facilitated merger is reflected in the current growth of cultural and academic interest in the cyborg—the human-machine or bi-
ological-technological synthesis-symbiosis theorized by Haraway (1985).[1]
"Increasingly mediated by computer technology...we are being...trans-
formed into cyborgian hybrids of technology and biology through our
ever-more-frequent interactions with machines, or with one another

through technological interfaces" (Dery 1993b, 564). With respect to VEs, if the cyborg hybrid "takes place" at the shifting focus of the point of stereoscopic vision that is always before us — and if therefore this cyborg is in and of the VEs on the other side of the screen immediately in front of the human eye, as well as in the user's consciousness — then the interactions or materializations Dery notes happen partly on a machine terrain. Stated otherwise, consciousness cannot be argued to predate the forms in which it is organized unless one seeks a metaphysical or idealist explanation. Experience in a VE feels less like Dery's conduit-implicit *through* and more like *on* or *in* technological interfaces.

Metaphors of discovery often veil strategies of invention, and those involved in writing and designing computer software and hardware think of these machines not as a discovery but as inventions (Bolter 1984, 75). They are positioned as conceived, then manufactured, not as found or perceived. Note the greater temporal dimension in conception and manufacture here, and the thoughtful planning it demands. This is so even when customizing a virtual world according to one's requirements, as figure 7 is intended to show. Here, the client, using the software and hardware provided by the VR manufacturer, customizes an application according to his specific needs. This is an act not of finding or direct perception but of premeditated design and integration of meaningful images

Figure 7. Client constructing the contents of a virtual environment. Copyright 1994, Division, Inc. Image courtesy of Division, Inc.

of nature, social relations, and agency. For the perceptive user in cyberspace, the quasi merger with someone else's conception, invention, or (pre)engineered meaning is not entirely dissimilar from aspects of everyday life, as when, for example, a teacher instructs her students in the ways of the world, or a skeptical worker maintains his instrumental acceptance of an employer's corporate politics for purposes of economic livelihood. Especially in a "world economy," there is increasing interdependence. Karl Marx's dictum that we make history but not entirely under conditions of our choosing retains salience and can be applied to virtual realms. Ironically, limitations imposed by the forms of virtual technology insert a contingency into what is argued will be the freedom to author environments at will. Limitation and contingency, the resistant materiality of places and lived worlds, echo in their ersatz counterparts, if for no other reason than that virtual technology's formal dependence on language, sight, and optics introduces meaningful limits and bounds—however much this may be ignored in the rush to virtual living by military men and thrill seekers alike. This limitation may produce unanticipated effects when it clashes with desiring expectations for absolute freedom in absolute space.

Body Language

> Travelers on ... virtual highways ... have ... at least one body too many—the one now largely sedentary carbon-based body at the control console that suffers hunger, corpulency, illness, old age, and ultimately death. The other body, a silicon-based surrogate jacked into immaterial realms of data, has superpowers, albeit virtually, and is immortal—or, rather, the chosen body, an electronic avatar "decoupled" from the physical body, is a program capable of enduring endless deaths. (Morse 1994, 157)

Human bodies form a basis for social relationships. The poststructural "social body" is determined by linguistic categories, yet while this social body "may be named as a theoretical space, it is frequently left uninvestigated. It is as if the body itself ... does not exist" (Shilling 1993, 72). In VEs, however, are users' bodies really dispensed with, "parked" somehow, or "collapsed"? Or do such metaphors mask a retheorization of the social body by academics and others more than they uncover any threat to natural bodies, or how always shifting forms of power are reconfigured by and in optical technologies? Although a VE, for example, minimizes ambulatory experience, users interacting with virtual technology nonetheless constitute material phenomena engaged in practices. Users wearing

HMDs confirm a sense that technologies such as VR are able to obtain a grip on human bodies.

We experience place as embodied human beings. This embodiment is situated somewhere along a continuum alternatively and confusingly conceived to exist between what are loosely identified as "nature and culture," or "culture and civilization," or even "nature and civilization." Embodiment can be considered either to help link a now-individuated sense of self to a wider community or, conversely, to contain this self "inside," apart from the broader sphere of social relations, which then comes to be conceived as an organism or entity somehow apart from individuals. Embodiment is a leaky concept; it suggests nature, culture, even civilization. "The body and its actions ... have a richly ambiguous social meaning. They can be made to emphasize perceived distinctions between nature or culture as the need arises, or to reconcile them" (Marvin 1988, 110).

David Levin asserts that a bodily nature never encountered except in a historical situation is one that denies our abilities to resist oppressive uses of history or even history as epistemology. Bodies resist history and do so in an inventive manner. For Levin, save for its extermination, culture and history can do nothing to the human body. Its physical reality and form resist the text of history, except for that part of history that would inscribe itself "biologically" over nature. This resistance is plausible, in part, because unless we are to believe that humanly produced representations somehow might have preceded the existence of the first human beings, human bodies were present on the earth before the first story being told and before its recording in any representational format. Such an understanding is somewhat taken on faith, as for Levin, "no *eidetic* intuition" is possible that would allow access to the nature of the human body apart from its involvement with history.[2]

Human bodies, therefore, are an intriguing pivot for theory, and it is difficult to imagine any geography that would matter without them. They straddle the dichotomy erected between nature and culture, their space both influenced by social relations and influencing what forms these social relations may take. The degree to which theorists have remained unwilling to look at our bodies as powerful means of countering the hegemonizing power released by the nature-culture dichotomy is perplexing. Incorporating human bodies into understandings of social relations allows a broader, more defensible, if continually shifting, material base from which theory might develop. The concept of *territory*, for example, works well to describe the external physical reality that results from our crafting of a place. The agency of both individuals and collectivities is easy to acknowledge in the act of making a territory, but the roles of the bodies of a territory's makers remain implicit and vague. Cultural studies approaches include "The Body" within their concerns, but too often this body is also the "site" of the individual. At the outset of *The Practice of Everyday Life*, de Certeau assures readers that his examination of mundane practices "does not imply a return to individuality" (1984, xi). There is a critical distinction between the practice of individualism and the fact of our discrete individual or particular forms as human beings. I wonder, however, if it is coincidence that the essentialized, singular, yet absolute Body surfaces as a subject for inquiry after theorization of the bourgeois or liberal individual as a worthy project is set aside. Moreover, whether the human body is alone or part of a group, it must not be theoretically essentialized, economically reduced only to a *site* for struggle, lest (conceptually), for example, theorists simultaneously were to disembody themselves to the point where they would have nothing left with which (or even *from* which) to wage such a struggle.[3] It is also germane to suggest the interplay between advanced capitalism and certain forms of academic inquiry. In 1997 I received via the Geo-Ethics listserv a final call seeking papers on "The Body," to be presented at the Institute of British Geographers' 1998 Annual Conference, under the auspices of the Social and Cultural Geography and the Population Geography Research Groups.[4] The opening sentences for the call read: "The Body. Is the body dead? Has it been 'done'?" While the overall tenor of the call is progressive and seeks to reenergize academic discussion of bodily issues, nonetheless, as a "hook" to attract interest, such phrases suggest that "the body" is already a shopworn, threadbare commodity, ready to be recycled, while "cutting-edge" academics move on to better and more "marginal" areas of inquiry than the now outmoded and therefore soon to be "discarded body." Such commodification resonates with the absence of bodies in virtual worlds — they are outdated Cartesian *automata* and hence are unwelcome except as moving pictures.

The body I am interested in, however, is not the obverse of the Cartesian mind or some prepackaged concept. Rather, bodies are particular *and* plural, have minds, spirits, and take place in an evolving fashion; this is a universal that has always been the case and will continue to remain so. Our bodies are where we locate individual difference; my body is here, and yours is there. Our bodies, however, also share operational similarities, which is why I sympathize with you when you are ill, as I recognize that my body can operate in a similar fashion. Nevertheless, it is the establishment and performance of spatial differences among bodies that works in unpredictable and relational ways with other bodies and with places. In material reality, the subject's embodied agency always locates a place from which resistance may proceed, from which ideas presented to her or him may first be tested and evaluated. In a VE, however, this place — the place of the subject's body — seems to lie physically "behind" the cyborg agent, who (or which) to date is made to take the ephemeral form of a subjectivity manifested as a cartographized point of view or avatar within invented computer language.[5]

Don Ihde (1991) proposes that computers have bodies and that we have failed to recognize this because "computer bodies" take different forms and are made of different substances than our own. Notwithstanding that it is Ihde making this argument *for* computers, and not computers making it themselves, his proposal may be extended to suggest that under the sign of an instrumental Cartesian rationality, the cyborg may recuperate — albeit in an *engineered* and *representational* form — an embodiment lost to modern subjectivity. Such an engineered embodiment would come at the cost of human conjoinment with machine. This may seem a high price to pay to subjects enjoying the experience of individual identity formation, and its attendant right and duty to bear the burden of creating meaning out of an examination and conceptualization of one's own experience. When the burden of responsibility starts to fracture the self, however, such a cost might seem less high.

Ihde (1991, 72) also speaks of machine intelligence as distinct from the human kind. More recently, Grosz (1994) writes of corporeal intelligence as a gathering together of bodily knowledge, mental activity, sensation, experience, memory, and agency. Corporeal intelligence suggests holistic synthesis, but it is not a mathematical totality. It is not as if the hand performs the same tasks as the eye, or that the taste of salt has the same *value* as the sound of an alarm bell. Rather, this intelligence draws from all facets of sensory experiences that take place through and in human bodies. Corporeal intelligence depends on the interrelationality of the senses. It is similar to the "expressive intelligibility" that Walter (1988) accords to the identity of place. The various facets of places and bodies can be specified as distinct both spatially and in their functions. Yet they also interdepend with one another in fluid ways. These facets — whether the speech lobes of the brain, the heart pumping blood to these lobes, the stone wall that holds an early frost at bay, movable outdoor furniture that can be repositioned along the sun's transit, or the palette of human ability gathered in a team of specialized surgeons performing delicate neurosurgery — are skilled or honed in certain functions or spheres of work or perception, but not in others. Placing these facets into discrete categories, therefore, is understandable, but useful only to the degree that this does not lead to the humpty-dumpty theorization of their mutual unintelligibility or separate absolute freedom. They all contribute their due to the whole, the entire result being that the organism, the body, the place continues to function and change in ways more or less supportive of the relational mutuality and difference of each constitutive facet, admitted inequalities and differences in power and function notwithstanding.

French performance artist Orlan, organizing her career around cosmetic surgery in her quest to call attention to what she terms "the problem of the body," laments that "the skin is deceiving. But that's all we are given in life. It's unfair because one never is what one has. We feel odd when we look at ourselves in the mirror. But by changing, one attempts to minimize those odd feelings." VR artist Nicole Stenger (1992, 52) poses the rhetorical (and incoherent) question "Isn't it exciting to live twice? To walk to a party with an accurate reconstruction of your entire body, flesh and bones, stored on a floppy disk in your pocket?" Orlan claims her project is to help future generations mentally prepare for the problem of the body and genetic engineering, yet her comments about unfairness, made during an interview in the pop culture film *Synthetic Pleasures* (1996), suggest her profound disaffection from her own corporeality. Similarly, Stenger's proposal to liberate the embodied imagination exemplifies a central image used in VR's promotion — an appeal to cultural longings to transcend the limits of the flesh. Her disembodied fantasy, however, ignores the panoptical financial and military applications of ITs more generally, which are about *constructing* profiles of the activity and consumption *habits* of individuals in order to produce reliable data profiles. Playing with flesh-as-data anticipates individuals self-identifying as data profiles. These profiles become commodities used to model and quantify business decisions, and other possibly more sinister cyberspatial forms of prescriptive or anterior surveillance. The incongruity between what is promised at Stenger's implicit scale of personal pleasure and what is planned and proposed at a global scale monitored by telepresence, virtual intelligence agents, or even the Army Research Lab killing ground described earlier, suggests the continued power of the myth that the sphere of politics and its control of human behavior is discrete from the application of technology to control nature and benefit humankind. But when the human body, like nature, becomes an object for control — as it is for Orlan, who attempts through repeated surgical interventions to make her body a fully controlled environment — because the spirit or mind or intellect is understood as distinct from any bodily referent, then it becomes harder to argue for a hard-bounded distinction between science and politics. As a technology of representation, VR is made to cut both ways: to free the individual spirit, in part by recourse to metaphors of subjective transcendence that occlude consideration of spirit's connection to matter, yet also to control by surveillance of performativity the spirit's individual manifestation in the body's power and difference. As I argued in chapter 4, in a VE, users participate in their own surveillance. They consent to have the computer monitor their physical position and stance as well as their subjective iconic identity extended or projected into virtual space.

Cyberspatial politics will not leave users untouched. Human bodies provide the technology with immanence, with indwelling, but because Western embodiment is infused with modernist understanding, the spatializing distinction established between subject and object — individual and society, inventor and technology, politics and science — permits a certain negation of human responsibility for creation of concepts and tools. This naturalizing cultural process allows a belief to take hold that technologies are akin to a "new nature." Stated otherwise, for many contemporary individuals, because technologies increasingly take the measure of nature, they are believed to be one and the same with it. The dazzling power[6] of representations of nature in a VE comes to seem equivalent to "the real thing."

VEs suggest that a communications environment has become a new site for ritual, but this time human bodies are absent even as they are visually represented. Communication is severed from body motility in favor of imaginative extensibility across space as an ecstatically disembodied and delocalized ritual of in*form*ation. In this respect, human geography's concern with the "collapse of space" can also be understood as part of a wider concern with the "collapse," denial, or extraction of natural value from nature, and the erosion of access to meaning and instruction contained within the myriad ways that humans are part of the natural world. This "collapse of space" has been met in symmetrical fashion by the creation of virtual worlds. Their makers are confident these simulations, via language practices voided of their embodied aspects, adequately represent the natural world that seems increasingly estranged from many people's experience. By virtue of the perceived need for such a representation, VEs also suggest the effective "death" or "collapse" of this natural world, bodies included.

Transcendence and Dazzlement

[In VEs] . . . the subject is dissolved in the swirls of cybernetic information, but is at the same time further empowered through an extension of motility and spatial possession. Here, then, are the paradoxically simultaneous experiences of death and immortality that are fundamental to religious practice. (Bukatman 1993, 295–96)

Our bodies are where we experience the intersection of our individuality and the cultural sphere. While subjectivity seems amenable to conceptual relocation to "transcendent sites" beyond the individual's physical body, his or her body is not transcendent. Virtual technologies encourage belief that they constitute a "transcendence machine" within which the imaginative self might escape its privatized physical anchor and live in an iconography of pleasure. The etymological roots of "transcendence" lie in surmounting, of going beyond ordinary limits, but I see nothing in this that necessitates surmounting the human body, any more than it might refer to surmounting the mind's imaginative limits. Leszek Kolakowski (1990, 118) argues that through communicating, one knows one exists, but that transcendence, as a necessary counterpoint or "ornament" to our existence, "never becomes real as a field of communication." Transcendence, for Kolakowski, is acknowledged, but its aesthetic dimension precludes it from being an end of communication in itself.

Consider, then, the transcendent moment not as a realized fantasy of "escape" but instead as an imaginative engagement that extends subjectivity beyond the self. Now, engagement means an entering into, or a pledging of the self in challenge. Transcendent activity on one's own always involves work. The struggle to achieve demands an engagement with the world to the degree that it implies a wager we place upon ourselves that we will surmount a variety of limits. Promoters of virtual technology such as Stenger, with her body doubling promise of living twice, offer a glimpse of transcendence that appeals to fantasy and is packaged as a form of psychic assuagement, yet she makes no mention of the effort that transcendence demands.

I understand the use of image to be part of the communicatory function of language and art, but we can never physically reside within image. It is here that Ong's (1977, 320–22) recollection that visual technologies also are art can enrich Kolakowski's understanding that transcendence as a necessary ornament never becomes real as a field of communication. When efforts to achieve transcendence succeed, we are engaged with the world and less constrained by Elias's modern wall of self-alienation. We are also less hobbled by the compensatory fantasies of extension that seem to offer the aestheticized promise of vaulting across alienation without really making a dent in its "structure."

Art and language have the power to affect us. Through media, we become aware of other ways of interpreting the world. At the very least, this opens us to sympathize with other views, experiences, and places. However, we do not physically reside *in* a particular form of art; our existence cannot be "located" in transcendence any more than the classical light and the truth it carried could be "localized" in transcendency (or movement). This does not preclude such a desire from being expressed as a metaphor that relies on an aesthetics divorced from ethics in order to argue the moral good of what is really a social incoherency.

To the degree that language facilitates communication and human affectivity, it might be argued that languages are how we transcend existence and achieve a relational humanity beyond mere subjectivity located on a grid. Yet to imagine that existence could somehow be (re)located in communication understood only as a transcendent act implies a taking leave of one's person, and by extension, a taking leave of the earth. Transcendence is an ideal we strive for but never achieve as an experience that could be represented fully, even as thought. The feeling of transcendence may be universal and akin to the *luminosity* of "direct perception." However, the process of its achievement is contextual, ongoing, and without a single path. Any transcendent moment never fully detaches from its origin in existence. To sustain a belief in becoming, we must call on a sense of the eternal, but this, in the limited, living way that we may know it, is a power always experienced in the place of the existing self.

The wish to transcend the limits of embodied reality via "living virtually" is connected to a pervasive, metaphoric, and now naturalized understanding of communications technologies as "extending" the self across or through space. This extension, conceptual as it might be, also implies movement on the part of the subject through space, but this has the effect of restricting the meaning of transcendence. Concepts such as McLuhan's (1964) notion of electronic media as "extensions" of humanity and that new media are nature have had great influence. Applied to VEs, if McLuhan's thesis is considered in the context of James J. Gibson's notion that we literally grasp our environment, one is left with the disarticulated oddness of "direct perception" as a grasping or seizing performed via optical extension. It is true that at the scale of our body, we reach out—most often with our hands—to grasp aspects of the living world around us. However, hand movements are not detached from the physical body. In a VE, the user may operate a handheld device to propel her point of view, yet it is the VE that is doing much of the "moving." Although a limited degree of grasping is permitted by applications using an interactive iconic hand that allows the illusion of physical engagement with the VE, the user's real hand, like a sign, often serves only to point in the direction her eye will fly. Focus proceeds noncommittally, and sight divorced from touch permits the percipient to believe that she remains free of causal involvement with objects (Jonas 1982, 142–48).

This confusion between bodily engagement and a sense of self-extension achieved through a partial handing over of agency to technology positioned as virtual nature is also a confusing or blurring of the scales of meaning existing between body and machine, tool and technology. In a VE, the scale of the body is written over by the scale of the machine. Because (in sophisticated applications) the user seems able to grasp objects arrayed in virtual space—in Matsushita's "kitchen world," she may open and shut kitchen cabinets and drawers and control the level of water in the sink by turning the faucet on and off—she may focus on her agency to use the technology as a tool and overlook the importance and degree of engagement she has ceded in order to interact with the machine representation. The Matsushita "kitchen world" is a world of pure geometry, color, and light. The represented body—both as extended point of view and as the iconic hand that grasps the virtual objects in the virtual space—seems to merge within an idealized world of visual practice descended from Euclidean geometry, metaphors of light, and a belief that the mind centers consciousness to the exclusion of the human body-as-container.

The transcendency of Platonic light and its withdrawal to a pure and otherworldly state—locality in transcendence—is mirrored in the contemporary virtual world. If, for Platonism, few were equal to the task of facing the dazzling otherworldly light head-on, and if this then required light's becoming a metaphor for salvation and immortality, the *paideia* (education of youth) on offer in the current generation of immersive VEs is a faulty lesson confirming truth's absence from the material plane—whether the plane be of "nature" or of human cultural collectivity as evinced in Carey's 1975 description of ritual communication noted in chapter 3.

It has long been understood that the gathering of information, along with experience of the world and coming to understand causes and effects, is necessary before one might attain knowledge and the synthesis of thought. Thought, in turn, has been held requisite for any eventual wisdom and ability to conceptualize. Increasingly, however, in telematic environments, data gathering and sorting through shards and "factoids" of information passes for the wisdom of experience. The meaning of the productive but difficult and time-consuming struggle to change through learning and education is eviscerated by implicit promises that easy consumption of images will be coeval with knowledge. It is ironic that the world of information, so dependent on metaphor for its affect, seems little concerned with how knowledge and, finally, wisdom might be produced and acquired, save to assert that technology itself generates knowledge. Information — a series of facts organized as rules and routines capable of being acted upon — is made to bear the conceptual beauty of Platonic forms (Heim 1993, 89). Further, though new technologies organize intelligence in new ways and offer new forms of access to information, when Jaron Lanier, discussed more extensively hereafter, promises that we will build houses in VEs without any prerequisite skills, he perpetrates a hoax on the meaning of work and the production of knowledge that is hidden behind cultural assumptions that consumption now offers the best vehicle for identity construction. Such a hoax perpetrates the false promise of easy enlightenment — that easy access to utopia is "just around the corner." Yet it is safe to say that Lanier has worked very hard on theorization, invention, and promotion of VEs. In pronouncing to the rest of us that in essence knowledge will be free, he disregards the meaning and value of his own demanding work practices in contributing to and directing the knowledge that has led to any fresh concepts he may now have to offer.

Razzle-Dazzle

As discussed in earlier chapters, direct perception is analogous to looking *into* the light. I have noted the physiological impossibility of ever becoming accustomed to the intensity of dazzlement's pure light. Its brilliant power also connects to a mystical ambiguity. To be dazzled is also to be bewildered, somewhat stupefied or stunned, even to have strayed off a path or direction. Virtual technology introduces dazzlement closeup. Users donning an HMD stare into an immersive virtual world composed of light. Classical thought understood light's transcendent power as divinely given. The moral right to relocate this power to human technology flows in part from the belief, beginning with Cicero and fully developed by Enlightenment philosophers, that humans have discovered how to access the interior light or Good housed deep "inside" our individuated subjectivities. Immersive VEs combine implicit Platonic and Neoplatonic spatial relationships between seer and light. They first position users "at the margin" of the field of view they illuminate. At this point, eyes glued to the interface,[7] the viewer wearing an HMD replicates the Neoplatonic truth seeker dazzled by looking *into* the light. However, this viewer is much *closer* to dazzlement's source: the HMD, acting as a conduit, replaces the space between the subject and the light. Transcendent light in a virtual world is thereby detached from the sun's dangerous brilliance. The Platonic realm of Ideal forms now not only "comes into view" but, by use of the (Neoplatonic) HMD, seems inhabitable as the Ideal made manifest.

In chapter 4, I noted Held's and Durlach's use of "world-as-display," in which users are both *in* and *on* the display. I also noted the move from the more fixed Neoplatonic relationship[8] established by looking into the light — one that extends through the design of the camera obscura, film, TV, and video — to one *in* which the subject is suffused by light from every direction yet is also *of* the light. The possibility of "VR enlightenment" is promoted by the cultural instruction contained in figure 8, which ironically suggests that subjective interiority can be confirmed, augmented, and enhanced through a symbiotic fusing of human and machine, and of nature and culture. The subject of the advertisement, within a "world-as-display," becomes both a source of his own inner light and a reflection of the emanating light of others "within" the virtual environment. The ad is a technological updating of Plato's cave, an instruction to seekers of enlightenment and transcendence that Ideal forms radiate from the Absolute One — VR technology itself. The ad suggests that one might liberate oneself from the "shadowy cave" of bounded, material reality by coming to understand that the sensory world only reflects Ideal forms that are accessible in VEs as representations in and of the light. The material world is but a mirror reflecting an ideal illusion. The subjective, cyborglike interiority of VEs suggests that the Ideal continues to reside at least partially within the user. The computer-generated interactive iconography that dazzles the user insinuates the opposite: that the dream state on view not only may be generated within the individual (the premise of interactivity between users, or users and machine)

Figure 8. Enhancing subjectivity: user wearing Virtual I-O's "I-glasses."
Reproduced with permission of I-O Display Systems.

as an individuated wish for merger with perfect forms, but might also
reflect divine will reformulated in the garb of a contemporary technological luminosity. The information on display is visual. Jonas's explanation that the eye ascribes an a priori power to images, which are perceived
to operate as a cotemporaneous manifold, offers a clue to understanding the naturalness of "direct perception" for VR users. They may forget
or choose to forget that they are interacting with simulations and may
be ignorant of, or choose to avoid dealing with, the consequences of acknowledging the degree to which their visual acuity has a cultural dimension and a history.[9]

I am not implying that users are entirely unaware of these processes or
that the latter could never apply to other forms of visual representation.
People still understand the implications of the spatial difference among
being the light, being in the light, and looking into it. But in desiring release from subjective meaning, a *part* of contemporary subjectivity seeks
dazzlement and seems to care increasingly less about the meaning of these
distinctions. In an act of disavowal, thought processes and judgment
may be set aside. Compared to theater, film, and TV, VR's greater sense
of immersion and interactivity, and its implication of users' bodies,

abets this disavowal or cultural amnesia. This disavowing comes from a split belief, as in "I know it's just a story... but nevertheless" (Morse 1994, 176)—what might also be termed a splitting of sensation from the conscious knowledge that one is interacting with representations.

A second, related explanation for users' willing engagement with VEs flows from the instability or fluidity of the meaning of icons and emblems. This instability, or the always already potential for overdeterminacy, is why sacred images—for example, Catholic holy cards of saints—often have textual accompaniment. Text is understood to stabilize or hierarchically direct in advance the interpretation or meaning that the viewer is expected to accord or extract from the image. Images are understood to be more directly accessible and thereby subject to a greater degree of personal interpretation, which has the potential to undermine official control. In a future generation of VEs—as evinced in the hopes of Stenger, Benedikt, Lanier and others—it may well be the instability of the image that attracts many users. This instability promises enhanced ego control through the user's ability to select how her or his emblematic icon, as well as other information she or he might broadcast and receive, would appear on the spatial display.[10] The user appears to exert a control over the language underpinning the VE, which in turn controls her or his icon once selected or designed. One chooses from a topology of symbols that operate within a finite frame, yet each has a potentially unique meaning to the user—and from moment to moment, use to use, and application to application as well. This bodes well for a creative expression of the self as iconography and poses few problems should one's goal be communication between oneself and the machine. It is more problematic in the context of meaningful communication between people.

Schizophrenic Utopias, Psychasthenic States

I think there has been a long-standing dream of rediscovering paradise, so that, if we have urban space which has separated us from nature, then our dream of technology will be about technology giving us a pristine natural environment within the city that can incorporate parks and greenery and oxygen.

Narrative voice-over from *Synthetic Pleasures* (1996)

Life on earth has always been very boring for me.... Hopefully [on other planets] there will be other people like me where everyone will be accepted no matter what they are.

Young gay man speaking to the camera, *Synthetic Pleasures* (1996)

Humans have always searched for utopias. How these have been imagined has varied across cultures, but an essential similarity found in all of them synthesizes "two antipodal images: the garden of innocence and the cosmos... [we seek]... for a point of equilibrium that is not of this world" (Tuan 1974, 248). Utopian spaces are often otherworldly "solutions" for those who find themselves in all too mundane social and material reality shot through with conflict. Plato wrote following the Peloponnesian War, a period of arbitrary injustice, famine, poverty, plague, and the despotic acts of the Thirty in Athens (Bengtson 1988, 155). Forcibly excluded from practical politics in Athens, Plato in his work turns away from political concerns toward a theorizing of ethical and cultural realms. "The Greek city-state had outlived itself: for precisely the best citizens it no longer had any room. The spirit consequently fled into the world of the unreal, into utopia, and in this realm Plato's writings on political theory, his *Republic* and *Laws*, gained eternal significance" (156).

At the beginning of the modern period, Hobbes's *Leviathan* posits incessant war. Like *Republic*, it stems from a period of great civil strife and was published during Oliver Cromwell's "reign" as lord protector of England. Hobbes can be read with hindsight as setting forth the conceptual framework within which the possibility of a modern bourgeois state can be theorized. For a free market to come into being, political freedom must be somewhat curtailed for all individuals engaged in accumulation and trade, lest in their competition they naturally come to blows and cause society to degenerate into continual strife (see Hartley 1992, 122–27). The interests of the individual are to be represented by the sovereign who rules over the artificial state, or Leviathan. This sovereign representation is to take the form of "Means and *Conduits*" (Hobbes 1985, 383–85), the forms of which remain underspecified in his work.[11] The need for a conduit metaphor is apparent in his positing of the inherently self-communicatory "present to oneself" relationship between Author and Agent, as discussed in the introduction, and diagrammed in figure 6.

Plato's "immortal realm" and Hobbes's allocation of representation to the form of the sovereign are different solutions to the problems issuing from troubled times. Fourth-century Greek philosophy manifests the political circumstances of the age. As Couch and Geer note,

> Plato and Aristotle, who could no longer look with pride on a great and free city, took refuge in cities of their own mental creation. In the *Republic* of Plato and in the *Politics* of Aristotle are found the origins of the philosophic and literary utopias of subsequent times, such as More's

Utopia, or Hobbes's *Leviathan*. The ideal state was a kind of literary creation that evolved out of the troubled and depressing conditions of the
age in which it was written. (1961, 489–90)

During the modern period, "where" a utopian ideal might take place
has been subject to rethinking. Desire for an evolutionary transformation
remains, but its location is shifting from "the journey into 'outer space'
from a dying planet to the virtual 'inner' space of the computer" (Morse
1994, 169). The synthesis of antipodal images noted by Tuan is also apparent in the attempt to fuse simultaneity and instantaneity through
the use of VEs. The effect of this is to produce an illusion of displacement and a "be there here" belief encouraged by telepresence, which
collapses real-time manipulation of virtual images of physically distant
material objects with the objects themselves. The cultural capital un-
derwriting any such illusory belief must connect directly to the power
of telepresence in VEs. Depending on the application, users can control
the virtual world inside the computer, but if telepresence or telerobotics
is the point of the application, then users achieve what Manovich (1995b)
terms an "anti-presence."[12]

This seeming compression empowers the social dynamic noted by
Arthur Kroker and Michael Weinstein (1994, 131): "You only know you
are (actually) here because you are (virtually) there." The power of telepresence notwithstanding, VR also becomes the immaterial realm that,
in offering a multiplicity of stages for the performance of multiple self-
identities, legitimates the ongoing sundering of self-identity. It reflects
and thereby confirms the naturalness of contemporary demands to adopt
plural roles as a precondition for making sense of social relations. Utopias
are partly premised on a transfer of power away from the here—from
the immediate, whether this be the era, one's body, or a set of individuals—to a separate agency or realm, whether this be a philosopher-king,
an absolute monarch, or the digital information space made to operate
as a site of redemption for a disarticulated self. To play with self-identity and space "therein," contemporary VR users implicitly cede embodied power to a belief that illuminated Information is King.

With respect to the recent manifestation of utopian and magical thinking expressed as a spatial conflation of the discrete meanings of "here and
there," the so-called "death of God" has dispersed sacred meaning, part
of which has come to reside within technology. God guaranteed absolute
representability and transparent, luminous truth. Today "we dwell not
in transcendental light but in the shadows of mediation and withdrawal"

(Ronell 1994, 284). I would note that this has left us with the access mechanisms to earlier gods—the metaphors—stripped of the immaterial destination to which they were intended to provide a conceptual link. We are left with the taxi but no destination to offer the driver—whether he or she is a scientist or philosopher—except to go round and round (or "flow" through) the circuits of electronically mediated communication, along the way arrogating the sphere of meaning to an objectivist "language" practice, which is becoming coldly autonomous and whose expressivity is separated from its root. In this schema, language is composed of arbitrary symbols that obtain meaning by corresponding directly to the objects in the world. Rational thought becomes an algorithmic manipulation of such symbols, and human embodiment has no significant bearing on the nature of meaning. A dual metaphysics operates: disembodied authors manipulate imaginary spaces already accorded a magical power.

The continued quest for technological progress is rooted partly in a belief in human perfectibility—that the light or moral core of goodness is within us, yet we must work to find it. Technology is applied to making this exalted task of getting in touch with the light less onerous, and this alone becomes adequate moral justification for the focus on virtual transcendence machines and the imaginary futures users may rehearse "therein." Despite their differences of scale, both tools and technology extend our grasp. Earlier tools such as the metaphorics, geometries, and other "visible instruments" that, for example, Roger Bacon believed would more fully reveal "the form of our truth" and "the spiritual and literal meaning of Scripture," were intended to access God. For later Enlightenment theorists postulating a light within, the fixed source of Absolute light above still was not fully extinguished. Although for many today the God behind this light is missing, lost, or "canceled out," optical technologies may be seen to offer a substitute: to allow individuals to communicate to one another as participants within an ideal sphere of continuously circulating communication.

The utopian exteriorization of identity by recourse to metaphors and technologies of spatial extension is hardly new. I concur partially with Carey (1975, 15) that thought is public "because it depends on a publicly available stock of symbols." In a sense, we are called out of our private selves and into a more "exterior" world of public social relations each time we think. However, I also agree that "being a *body* constitutes the principle behind our separateness from one another and behind our personal presence" (Heim 1993, 100). An immersive VE experience or a text-based Internet discussion group where one may choose which gender to "assume" confers a certain (elite) equality (assuming access), but the quality of the encounter narrows because we act out only what we want of our selves. The reality of self-identities staged in a VE can be almost entirely socially constructed — a mix and match of institutional facts that promotes fantasy, ignores body biology, and eschews grounding of other forms of human agency than the social. Without meeting others face-to-face, "ethical awareness shrinks and rudeness enters" (102). The simultaneous information available through networks undermines belief that the real world is worth knowing, because the speed of such simultaneity suggests that everything can be represented. In contrast, "real space" is important because it contains or "leaves room for" the unknown, and "real-world resistance has made us develop a mind-set that contemplates, reflects, and mentally digests matters by chewing them over slowly and thoughtfully" (145). This is why Heim also notes the ethical reality of face-to-face contact and its ongoing erosion within a dematerializing "public sphere."

Given proliferating and plural self-identities, why predicate the discursive space of future VEs on the strict one-to-one subject-object relationship on display within current models? In chapters 1 and 4, I noted NASA's project to simultaneously represent plural icons of the self within a VE. NASA is building a split subjectivity machine, one that also situates *both* the viewer and the visual field or ground in such a way as to break down distinctions between them.

Jonathan Steuer defines VR as "a real or simulated environment in which a perceiver experiences *telepresence*" (1992, 77; emphasis added). Steuer defines telepresence as "the experience of presence in an environment by means of a communication medium" (76), although Mark Pesce, also a VR entrepreneur, asserts that telepresence is "electronically-mediated schizophrenia" (1993, 9), which could lead to "telepathology." Research with more immediate commercial potential than NASA's proposes on-line VEs in which individuals will assign different icons or semiautonomous drones of the virtual self to "travel" to different sites in the performance of their jobs. These individuals will act as editors or authors of what I term their own plural schizophrenia. In chapter 1, I cited computer scientist Gelernter's proposal that "software agents" will act as surrogate characters for ourselves as they travel along the spatially distant information byways of cyberspace. These "agents" are "smart" programs able to rescript themselves in response to the on-site learning they achieve in doing "our" bidding. Such smartness attributes to these programs a form of agency previously considered to be the outcome of human intelligence. Software agents are "self-reconfiguring" and are conceived to mutate, perhaps radically, in response to the different demands placed on them over time. Software agents are Gelernter's reasonable solution to "navigation" within incredibly complex, rhizomelike on-line environments. Freed from bodily constraints, there would seem to be no end in sight for the potential to fracture self-identity into fragments intended to travel to different locations of information, whether these be within one's personal computer, or data banks around the globe.

Brian Gardner, whose extension of James J. Gibson's theories formed a basis from which I critiqued direct perception in chapter 5, theorizes "software agents" and plural self-identity as follows:

> A major problem with existing artificial worlds is that they can only be experienced from one perspective at a time. . . . One . . . solution is the idea of "invisible cameramen." . . . think of these as people who travel through the environment that has been created and try to analyze it based on what the user is looking for. (1993, 102)

Working similar code-dependent terrain, VR theorist Brenda Laurel, who believes metaphors of theater best convey the potential of virtuality, prefers naming such cyborg agents *characters,* not people. In her implicit neo-Hobbesian schema, people are central controllers (Hobbes's Author), directing the actions of a variety of representational subidentities (Actors) in pursuit of multiple goals. However, the relationship between author and actor is more tenuous. Virtual agency "that is responsive to the goals, needs or preferences of a person — and especially an agent that can 'learn' to adapt its behaviors and traits to the person and the unfolding action — could be said to be 'codesigned' by the person and the system" (Laurel 1991, 148). Within VEs, the various images of agency and person that underpin "cameramen as people" and "agents as characters" perform an iconic return to what Williams (1983, 233) defines as *person's* "earliest meaning of a mask used by a player." Hobbes notes that the word "*Persona* in latine signifies the *disguise,* or outward *appearance* of a man, counterfeited on the Stage; and sometimes more particularly that part of it, which disguiseth the face, as Mask or Visard" (1985, 217). As with the Greek masks of tragedy and comedy, the iconic "person" or puppet within a VE is an ersatz geographic individual on its way to becoming a technology. The Person-as-Actor is akin to Actor-as-Picture and moves within a "personalized" spatial display of visibilized language written by the Author. The Actor becomes the talking picture at a distance for the Author or ventriloquist, who writes the words the machine will translate into images the Actor will perform or become.

To permit access to VEs, immersive virtual technology masks the user's eyes behind an HMD. For the ancient Greeks, masks permitted a transcendent glimpse of cosmic mysteries. They allowed access to difficult truths, and to a way of coping with extremes of the human condition. A second skin, they facilitated staring directly into the face of horror or tragedy when mere humans would have needed to avert their gaze. As such, they can be theorized as a cultural technology that bolsters self-confidence, thereby also addressing the Platonic moral concern that few humans are up to the task of facing the intensity of dazzlement and enlightenment head-on. Immersive VEs require that users turn their eyes and backs away from the real world as a precondition for participating within digital sensation. While cyberspatial engineers work toward a future in which shape-acquisition input devices dispense with currently cumbersome exoskeletal costuming, for the moment, variations on the DataGlove and DataHood, and even virtual image displays, remain the contemporary immersing equivalents of this ancient Greek interfacing device. They access "the ideal" in the form of an imaginary territory that *seems* an extensive and dazzling facsimile of cosmic power.

Whether or not future information-gathering scenarios will feature a host of invisible (or "masked") cameramen or characters doing part of their masters' bidding, as Laurel implies, the individuals on this side of the interface will have consented to transfer a degree of agency to these devices. This is not entirely a change in kind from what exists today, given an increasing cultural reliance on all manner of external "memory" technologies such as computer hard drives, voice mail, answering machines, and Web "push" technologies; spatially extending devices such as the simple telephone; or even such basic servomechanisms as photoelectrically operated doors. However, this next step in the exportation of agency from our embodied self would represent at least a change in degree. It is germane to ask at what point the cumulative effects of changes in degree become a change in kind. Given our immersion in places and apparent need for cultural forgetting of our agency in the making of technological change, would it even be possible to identify with any precision that a change in kind had actually occurred without, for example, the benefit of "hindsight" or history, whose linear narrative these technologies have the tendency to unsettle or even occlude?

With respect to the possibility of a change in kind, cyberspace will not only be populated by dronelike simulations of information-seeking quasi-robotic aspects of the self. As is already evident with contemporary on-line text-based environments and discussion or "chat" rooms, cyberspace will likely be graced by the multiple presences of individuals assuming the "playful poses" of various identities. This much-lauded "playfulness" seems an acting out of the plural identities that contemporary individuals assume as part of the demands placed on them by what is variously termed postmodern culture, late capitalism, or globalization. However the current state of affairs is conceived, increasing numbers of individuals do experience a sense of fractured self-identity in the real world — now a mother, this evening a consumer; now a writer, tomorrow unemployed; now Caucasian, now a student, now the adult child of an alcoholic, and so it goes. Both stress-inducing and liberatory, this kind of self-conceptualization and the pluralistic identity politics and consumer lifestyle it promotes seems natural to large numbers of Westerners. However, if in a VE I cannot confirm my self-identity through relying on your reaction to me because today I am adopting a different persona than yesterday, then *who cares* when identity has been reduced to a flow of signifiers? Representation is only so elastic. When texts and icons start to represent themselves, the centered subject is threatened, for it becomes difficult for others to identify with him or her through the use of these referentless symbols that direct attention away from their representational function. If place is "a field of care" — as a symbolic topology or field of ideal illusions, a VE could only ever attain a partial quality of place. The cyborg actors on the virtual stage are at once in and out of place. The split between a physical body in the real world and a phenomenal body in the VE may suggest that the body is only "a site for departure and return" (Friedberg 1993, 143) of a too-fluid self-identity that moves back and forth between body and referents as if all were nodes within a communication system.

It seems unexceptional that such a transference of identity is argued to be available by leaping through the screen into on-line environments. These technologies are coming into production precisely to cater to the demands of plural identities. What does seem different with VEs is the full flight of fantasy that can be achieved "therein." It seems a change in kind that on-line "invisible cameramen" will continuously circulate at the same time as different cotemporal on-line manifestations of the self seek out an ersatz merger of stage and experiences and/or new information.

If realized, this change in kind would imply an unthinking acceptance of schizophrenia as a natural way to organize social relations. Schizophrenia is, in part, a profound disturbance of the self's relationship to space. Schizophrenics often view the world as an extension of themselves and at their whim, and they may be afflicted by a loss of identification with their own experiences. They report that their thoughts exert magical influence on material realities, and they may think they are identical with external objects. Yet they also relate that their thoughts are not thought by them, or that things are not seen by them, but only by their eyes. Common objects can seem like exteriorized projections of one's mind that paradoxically doubts its own existence — a blending of loss of self and unrestrained indulgence in hubris (Sass 1994, 86).

"Perhaps the most emblematic delusion of this illness is of being a sort of God-machine, a kind of all-seeing, all-constituting camera eye or copying mechanism ... an ... ultimately self-deceiving preoccupation with, and overvaluing of, the phenomenon of [one's own] consciousness" (87). Louis Sass recounts Victor Tausk's influential 1919 essay "On the Origin of the 'Influencing Machine' in Schizophrenia." Tausk's patient Natalija believed herself under the influence of a spatially distant machine operated by unknown people. Yet this device had the "form of her own body and everything that happened to it happened also to her." Natalija was, according to Tausk, "picture-seeing in places" and attributed this to an influencing machine such as a "magic lantern, cinematograph, or other representational device" (Sass 1994, 92).[13] Natalija's testimony calls attention to a technical apparatus. VEs as well as the cinema rely on what Jean-Louis Beaudry (1974–1975) refers to as a "persistence" of vision that allows the technical apparatus of a visual technology to be forgotten (42). Beaudry's point about vision complements Jonas's (1982, 136) earlier-noted assertion that sight establishes a cotemporaneous manifold and requires no perceptible activity, thereby withholding the experience of causality (149). Beaudry asserts that film frees the eye from bodily movement, instead constituting the world for the eye and along the way making invisible both the apparatus of film and the ideologies at work (1974–1975, 43). As such, the camera's movement somewhat corresponds to the "transcendental subject" that Natalija at least partially believes herself to have become; she has become less a subject and more a picture to herself. VEs, however, do partially rely on bodily movements as part of how they then visually simulate a world. Unlike cinema, they introduce multiple perspectives — including the ability to look at external representations of oneself as a separate icon or object in space, as shown in figure 6. VEs suggest the subject's involvement in the shifting space, and more so than with cinema, in a VE, "I *am* a Camera" — a kind of avant-garde positionality replete with multiple perspectives and lenses. This is a change that corresponds to what I understand as the retheorization of schizophrenia in still more spatial terms — the *psychasthenic* condition theorized by Roger Caillois (1984).

The works of Caillois and Celeste Olalquiaga (1992) are useful in considering the dynamics of any weakening of spatial distinction between self and world. Extending Pierre Janet's earlier theory of *psychasthenia* as a draining away of energy from the self (Starobinski 1989, 150), both authors identify a psychasthenic merger of self and surrounding space that results from confusing the physical space of a person's body with (its image in) represented space. Caillois (1984, 28) identified this as "a disturbance in the perception of space." His observations, published in 1937, have been extended by Olalquiaga, who, writing about the "vanishing body," notes:

> Confused by transparent and repetitive spatial boundaries, disconnected from the body by a video landscape that has *stolen its image*... contemporary identity... can opt for a psychasthenic dissolution into space... floating in the complete freedom of unrootedness; lacking a body, identity then affixes itself to any scenario like a transitory and discardable costume. (Olalquiaga 1992, 17; emphasis added)

Caillois treats a psychasthenic manifestation as a failure of the subject to achieve a unification between body and self. This failure permits space to act as a "lure" (see Grosz 1995, 89), drawing identity outward in the absence of achieving self-location in one's own body. Coherent identity devolves from an embodied subjectivity whose position organizes how the subject sees the world around herself or himself. Caillois — putting words into the mouth of a psychasthenic individual — writes, "I know where I am but I do not feel as though I'm at the spot where I find myself" (1984, 30). In other words, the self seems a nonself to itself. Caillois's succinct phrasing prefigures Kroker and Weinstein's (1994) earlier-noted description of VR as an incoherent strategy of knowing that one is "here" (that one *exists*) because one receives self-confirmation by being (virtually) "there"; that is to say, by *seeming* to exist in a pastiche that cobbles together the reality of one's body with a set of surficial images constructed beyond one's bodily coordinates. A decentered contemporary subjectivity in turn constructs a cultural technology that *reflects* this fragmentation. Like Narcissus, this subjectivity mistakes reflection for its own embodiment.

As a concept, psychasthenia is an explanation of schizophrenia as spatial disturbance between the self and the living world around it. VEs simulate a psychasthenic merger of identity with represented space. This is not the same as the relationality between people and places, or even any sympathetic or empathetic interdependency between people and their environments through which, for example, other people form part of my environment and I am a part of theirs. Rather, in psychasthenic fashion, VEs herald a perceptual fusion between self and surroundings by suggesting to users that surrounding space, as if by magic, need no longer materially resist their willful actions. In this way, VEs may legitimate schizophrenia as a model of identity and social relations.

The phenomenon of "apart/together," which characterizes the alienated world of consumption, leads to increasingly fantastic juxtapositions (Sack 1988, 657). VEs suggest the eventual promise of a phantastic merger of self with space, identity with machine, agency with commodity, flesh with silicon, reality with fantasy, and existence with communication, this last conflation achieved only and entirely in the metaphor of "information space." Recalling Rothenberg's 1993 circle of technology by which intentions get transformed into technologies that in turn suggest new intentions, a psychasthenic confusion between bodies and their representational images might lead to an "intention" to craft a technology to "more adequately" represent this confusion. As such, VEs give instance to abandonment of one's own embodied identity in order to embrace space beyond the body. At least part of identity would then reside exterior to the self. Many technologies promote this exteriorization,[14] though Olalquiaga (1992, 2) relates psychasthenia's emergence to the increasing social importance of visual images in the production of simulated spaces such as TV. What may differ with virtual technology is a loss of existential orientation. If spatial and identity polyvalency are to be the pluralist "norms" in cyberspace, a resulting sense of unreality may promote extreme disorientation. However we conceptualize space, we each have some personal grasp of what it is to be "lost in space." What might it mean to be lost in ersatz reality—disoriented within a "spatial metaphor" powerful enough to suggest the naturalness or historical inevitability of the psychasthenic mode of experiencing—a mode Caillois identifies as a social pathology? As Olalquiaga suggests, when representative images replace symbols in the communication of meaning, access to abstraction is reduced (4), and with it an ability to distinguish critically between fact and fiction, or to sense the impossibility of residing in the "visual space" of a metaphor. This is akin to believing one could live in, or more precisely coexist *with*, the "space" of a dream. We are unable to conceive of existence except in space (Jones 1982, 69). However, this is not to suggest any merger of self and surroundings. Jones notes that "to exist" is derived from the Latin "to stand out" or "to emerge," and what we stand out from is space. This is quite distinct from the premise that in a VE users form part of the display.

I close this section by returning briefly to the distinction between changes in degree and kind. It seems a change in kind actively to seek conjoinment with and within a technology that enhances a conscious experience bearing so many of the hallmarks of schizophrenia. This is a consequence of belief that our consciousness and its need to communicate its own ephemeral immateriality somehow is suzerain over material existence itself. It comes to pass that the spatialization of this consciousness, achieved by virtue of its need to communicate itself to prove that it exists, minimizes the environmental evidence of the places of the earth whose resistant materiality confirms their inhabitants' existence. Existence presses back on us; it does not permit us to achieve ever greater polyvalent identities without cost. The sanity-rooting constraint of real places allows us to dwell and interrelate with the phenomena around us with a fair degree of certainty that these phenomena, while evolving over time, will maintain their own separate and relational continuities. It is partly through recognition of these continuities — exactly what promoters of VEs suggest that users cast aside — that we achieve external confirmation of our individual and group identities.

Monstrous Spaces, Desirable Visions

I am concerned with how virtual technologies might foster a myopic withering of social identification with real places, lived bodies, and the nonhuman parts of material reality. In this regard, I note Haraway's 1991 consideration of the possibilities for vision and its engagement with nonhuman forms of agency. She argues for an embodied reappropriation of our most powerful sense. "Eyes have been used to . . . distance the knowing subject from everybody and everything in the interests of unfettered power . . . visualizing technologies are without apparent limit" (188). An extended but still embodied perspective "gives way to infinitely mobile vision, which no longer seems just mythically about the god-trick of seeing everything from nowhere, but to have put the myth into ordinary practice" (189). Yet Haraway refuses to cede vision, at least metaphorically, to the forces behind this myth. Her eye pleases her, and she argues for an "embodied objectivity," necessarily partial, as a tool for making common political cause against the god-trick "in order to name where we are and are not" (190).

Haraway's "embodied objectivity" is an effort to move away from the overly ideologized subject to find a common ground for science and nature. She criticizes how vision has been deployed within unequal power relations, but she is hopeful for its eventual redemption. Yet the difficulty in relying on sight alone to make such a move lies precisely in the eye's concrete abilities. Both the powers and deficiencies of the eye are asymmetrically prescribed and limited by the body of which they are part. In allowing that embodied objectivity may finally be contained within a high degree of metaphor, Haraway may inadvertently aestheticize the currently overly subjected feminine object she would rescue from "the gaze." By a roundabout manner, she may arrive at a "nature" that, in Cavell's phrase, is little more than "a world viewed," with all the attendant alienation from same. I find this frustrating in its implications that fully cosmopolitan eventualities, and the mutating technopolis that undergirds their structure, are the only viable or possible futures before us — that the most agents can do with the products placed before them by Nintendo, Sega, or the Pentagon is to interact creatively with them under the guise of resistance. It is unclear that such resistance can remain immune to technical impingement.

I do agree with Bruce Mazlish, who in 1967 had already concluded that refusal to engage creatively with the change to our metaphysical awareness being wrought by the erasure of difference between humans and machines leaves us with the unpalatable alternatives of frightened rejection or blind belief in machines' ability to solve all problems (1967, 15). As he notes, this Hobson's choice is in part what gives Mary Shelley's *Frankenstein* its ongoing appeal. Examining how differing forms of technology intersect to inflect our "metaphysical awareness" remains essential, even if such work is labeled reactionary by those who argue that this kind of examination depends partially on a politics of the natural that was always only a cultural artifact anyway.

There are *many* ways to proceed. As a culture and as individuals, we must acknowledge some responsibility for the actions or effects of technologies in use and not bury them under the rubric of a passive and unacknowledged belief in technological progress operating as a kind of contemporary providential metaphysics of vision. Privileged elites have written most of the West's philosophy, set the machine agenda, and run the world. What kind of seeing machines might be designed by others vying for fuller voice, given that people worldwide still struggle to obtain an increased measure of the subjectivity that privileged Western elites would so readily abandon to the cyborg "positionality" of VEs? Is the technical trajectory of virtual technologies, which promise relationality while depending on absolutes, already cast? *Seemingly* so. But might it be broken? It must remain intellectually acceptable to argue for (a part of which is to philosophize) different technologies than the dominant model of the cyborg now under construction — a too accommodationist agenda that confuses visual metaphors with physical bodies, and that fuses electricity and flesh, number and light, within the disciplinary grid of an already imbalanced equation. I make this argument recognizing that any such future and ongoing attempts at social responsibility must begin within a context of unequal power relationships influencing whose intentions, and therefore which philosophies, ideologies, and discourses, get transformed into technology and naturalized.

Almost always, it is elite intentions that get transformed into technologies. Fantasy is used to suggest the universality of elite philosophies and fantastical desires, even though VR's real power "resides with those who build the systems, design the software, and decide who is allowed to use it" (Friedberg 1993, 145). Consider the following example. Lanier argues that concrete reality constrains the imagination in pernicious ways. Think, he suggests, of how difficult it is to build a house. In VEs this will no longer be the case. The frustrations that attend embodied reality will give way to the imaginative design capacities of "everyman" transformed into a virtual carpenter, whose pleasure in building his virtual house will be courtesy of the requisite software. Lanier is intrigued by the "pliancy" of VEs, suggesting that what is striking about such worlds is that distinctions between human bodies and the rest of the world are "slippery" (Lanier and Biocca 1992, 162). Yet he is able to suggest, in almost the next thought, that in VEs, "the objective world is completely defined . . . therefore, the subjective world is whatever else there is. Suddenly, there's a clear boundary for the first time" (163). On which side of this visually dependent distinction that might leave our "slippery" bodies and any "pliant agency" remains unclear. Lanier suggests that virtual house building will be part of a (future) communication without codes. "If you make a house in virtual reality . . . you have not created a symbol for a house or a code for a house. You've actually made a house. It's that direct creation of reality; that's what I call post-symbolic communication" (Lanier and Biocca 1992, 161). Proposals such as Lanier's flow from the American virtual research community's embrace of James J. Gibson's absolutist and culture-effacing "ecological" theory of an evolution-dependent, hardwired direct perception of texture gradients and perceptual invariants. They seem a conjuring[15] that denies not only the interplay between culture and its symbols and the physiology of sight but also the hard-won lessons that building has to offer, not the least of which are those learned about the relationship between bodily limits and creative reach. Writing of his imaginative engagement with a work of art, Igor Stravinsky notes the power contained within the necessity of limits:

> In art, as in everything else, *one can build only upon a resisting foundation.* . . . My freedom thus consists in my moving about within the narrow frame that I have assigned myself.[16]

Both Lanier's belief in a future codeless communication and the correspondence approach to VR of which it forms a part are indebted to an understanding that within Western representational art and iconography, the space of a picture is formal but not codified, even though specific standards inform the maker's work (Bolter 1991, 53). As such, a picture makes a stronger claim to *reflect* the visible world. However, even Gibson's texture gradients and perceptual invariants can be understood as forms of coding. Modern information theory understands a code as a "set of statistical rules, a form of stored information" (Campbell 1982, 256).

Lanier's pacifying proposals recall Marie Antoinette's lifestyle at Versailles and a more general aristocratic donning of the appearance of workers' labor as leisure. We may rest assured that Lanier's proposals will do little to alleviate housing shortages, substandard accommodation, or homelessness. If man does not live by bread alone, even Marie Antoinette's pitiless cake seems preferable to Lanier's quasi-alchemical, pliant, and homeless image. I must acknowledge, however, that VR may also be experienced as providing a sense of bodily holism that might virtually suture the Cartesian mind-body split working against recognizing the bodily interdependencies I discussed earlier in this chapter. Mind-body dualism reifies the thinking individual or intellectual or elite versus the laboring body or mass. But psychically this is unsustainable, and Marie Antoinette was only one in a constellation of well-heeled patrons that has played at stitching the split. VR technology can seem to offer a reuniting of hand and head — a "therapeutic" environment wherein bodily motility and vision-as-insight are reunited through the machine.

Lanier intends his vision to be liberatory. He fails, however, to address the issue that VR trades in *images* of real things, as well as in simulations with no referents. Bolter (1991, 225) writes that the desire to break through to an immediate perception of reality reflects a deeply rooted Western longing to cut through the abstract networks of signs — the decoding of which print culture demands — and to deny the value of semiotic thought. The irony of Lanier's expectation that the iconographics of immersive VR might sustain an immediate perception of reality seems almost palpable. The antecedents to Lanier's vision seem more rooted in exploitation than liberation, and they depend more on semiotics and signs than on any direct connection to an unmediated reality. Lanier himself has forgotten the screen, if not the frame. Lord Castelreagh's scheme to export London's poor to overseas colonies is an earlier cultural technique somewhat precursive to Lanier's vision. The downtrodden were offered a utopian image of a "new world" — a colonial landscape of plenty and ease, almost as if the necessities of life would present themselves as commodities born without the intercession of labor. This was "aristocratic control over representation *for* the working class" (Bunn 1994, 136), in which the to-be-expected interplay between labor, land, and social relations was masked behind a description of the fantastic. The use of aesthetics to organize these kinds of utopian appeals is intended to transcend the particular conditions under which life is to be lived (Bourdieu 1984). The exportation of Victorian England's underclasses partially depended on an appeal to fantasy (not coercive, but dazzling) designed to deflect emigrants' attention from the conditions of production that would await them in distant "utopias." A fantasy image, based on a deceitful appropriation of nature, of a carefree, bountiful world of ease avoided issues such as native populations, unexpected climatic extremes, infertile soils, and social dislocations: in short, the to-be-expected intersection of the natural and social worlds. The kind of easy-to-assemble, do-it-yourself landscape that dances before Lanier's eyes is also a magical forward vision or advance brigade. The spatial "prospect" opened here is a projection of future development. It seems a paltry and enfeebling vision, but perhaps this relates to a nostalgic role of landscape as a way for Lanier to reimagine and project forward an earlier time than recession-whipped California in 1992 — to a time when the destiny of metropolitan cultures seemed "an unbounded 'prospect' of endless appropriation and conquest" (W. Mitchell 1994, 20).[17]

Lanier's vision of a virtual house and the cozy comforts of hearth and accomplishment has a separate and earlier antecedent in the Russian historical myth recounted by Manovich (1995b). To accommodate Catherine the Great's wish to see for herself rural peasant living conditions, Potemkin, her first minister, ordered a series of facades of villages be built along the route of the regal visit. Like the trees at a distance in the military VE described in the introduction, the ersatz villages were constructed at a considerable remove from Catherine's view. The great empress, who never left her carriage, was left with the impression that her miserable subjects enjoyed a thriving prosperity. Like VR, the illusions of Castelreagh and Potemkin are all codes, and the viewer fills in the blanks that also serve to complete the poignant illusion or, depending on one's point of view, lie.

A related criticism of Lanier's "communication without codes" or "post-symbolic communication" draws from Searle's 1995 distinction between brute and institutional facts, developed as part of his discussion of how aspects of reality are socially produced. Searle describes a common misunderstanding about how language works in the following way. Words and expressions have senses and meanings, and therefore referents. This leads to the assumption that if we can think the sense of something without words, we can also think the referent without words, and that we do this merely by detaching the sense or meaning from the expression in words and just thinking it (67). That we can translate expressions of meanings into other languages, for example, appears to confirm that we possess a "thinkable sense" separable from language. This may be so for how we sense brute facts such as feeling the rain fall on our face. It is not the case with institutional facts. Searle uses the example of sports fans watching a player score a goal in a game. They see the reality of the individual crossing the goal line. However, they do not "see" the player "score six points." These points are not "out there" in existential reality as are the brute facts of the player, the ball, and the line. Scoring six points is a socially constructed institutional fact that depends on language for its genesis and meaning (68). Lanier avoids dealing with the acculturation and learning required for making sense of institutional facts. This leads him to make the implausible suggestion that mediated communication might exist independently of representation. The desire to escape the dualism between communication and existence, the representation and reality that underscore the utopian thrust of cyberspace's promoters, is at its core ironic (Markley 1996a). This thrust imagines cyberspace as a visual language, that is "the signature—not the sign—of a mystical unity between the inscription and the perfect intelligence behind the inscription" (65). The mathematics of form that cyberspace's engineers have chosen as the vehicle by which to move beyond representation "is rooted in the very binary thinking that it seeks to overcome" (65). Now, binary thinking requires the use of Searle's institutional facts. As in the idea of a double negative canceling itself, a representation is used to overcome the limits of representation in the hopes that a utopian existential moment might be arrived at as the result of a quasi-magical cyberspatial agency operating as an unacknowledged exodus machine.

The current popularity of virtual technologies partly reflects the technical elite's hope that these machines might represent, in commodity form, an acceptable commons in which fragmented, "cocooning," but highly individuated modernist subjectivities might achieve virtual reunification with other such selves without having to venture from behind real-life "spatial walls" such as gated communities. In this digital commons, plural identities in flux, feeling overburdened by the weight of their own subjective freedom, might be *pictured* more as one in their search for a transcendent continuity — as if returned to a prelapsarian state. Hence such prophecies as Kelly's hive mind.[18] Following the discussion of psychasthenia, I can now describe hive mind as evincing a desire to become space itself, or to invent a space one might occupy, but which would in turn replace the subject. Kelly's fantasy of animating information technology reminds me of Roman Polanski's intriguing cinematic treatment of schizophrenia, *Repulsion* (1965). Catherine Deneuve relates to space as a devouring force — the walls of her London flat reach out to grab her body as she flows trancelike yet terrified along its interminable corridors. This allows her body to dissolve into the transcendental realm of her fantasies and thought, in which the meaningful reality of the space around her seems both somehow doubled and reduced to Thanatos-as-space. The figure-ground relationship that helps root modern subjectivity is sundered and replaced by what Rosalind Krauss (1994, 155) theorizes as a ground on ground. Kelly's prophecy predicts empowerment achieved through the ritualistic gathering of on-line fragments of the self. This gain Kelly foresees, however, is to be made by ceding partial control to a machine premised on undoing the geography of human form.

A virtual on-line commons, like absolute space, would offer an infinitely extensible grid for potential reunification of fragmented selves, with plenty of room for commerce, too. Such a digital "public sphere" might permit imaginative vaulting of Elias's modern wall. However, the critical spatial separation of users' bodies in "absolute space" would remain unaltered. Each spatially isolated individual would become, as it were, a discrete modern category unto himself or herself. VEs — a vision disguised as space — are the Ideal public sphere for imaginative subjectivities believing "themselves" virtually freed of bodily constraints. These environments are a privileged psychic variation on contemporary physical homelessness. Divorced from the body's constraining intelligence, "the fully extensible self" — assuming sufficient personal wealth — busies itself with the shallow fantasy of building a virtual house in a postsymbolic "environment." At the same time, this self ecstatically embraces the digital means that serve to extend its own psychic disarticulation, hoping to sharpen its ability to control personal meaning by consuming virtual experience as both leisure and culture. Along the way, it turns its "back" on an intransigent material homelessness embedded in grossly unequal social relations.

Home Base

Our bodies are necessary for making the judgments upon which decisions and agency are based. If space can be thought to organize relationships between things, it also contributes to regulating human perceptions, in that they appear to occupy the same perceptual field. "This perspective has no other location than that given by the body" (Grosz 1994, 90). We grasp space through bodily situation.

At the very least, the strategic use of space within VEs will require rethinking. Apart from concerns about bodies, what is history in a VE? If an individual is to have the freedom in play or requirement at work to don many different identities, what will history mean, given that these multiple partial identities themselves are idealized machine-dependent hybrids composed of icons, simulations, and a merger of others' concepts with one's own sensation? If, in a VE, our physical bodies seem in the way of "progress," then it becomes easier to accept that the kinds of intelligent agents and cyborg characters Laurel (1991) proposes might really be more advanced or "evolved" than our embodied (and hence, by this logic, outmoded) selves. This thought can also be inverted to suggest that VEs confirm that becoming more fully human requires more technology. To this should be connected the possibility that, in a real way, VEs remove even the spatial ground of resistance that human bodies have offered against the misuse of history. They re-create a simulation of space that, potentially, is completely open to surveillance, transcription, review, and censorship. Any history on view, and the possibility to root identity in real places, becomes less important than the continuous circulation of hybrid discourses that masquerade as space.

There is a widespread acceptance within contemporary academic theory of the Foucauldian notion that power is weakest at the moment of its application, as this is when it is most visible. Whereas military stealth technology, for example, seems predicated on this understanding, telematics may suggest a new working for power. If these networks are where power now *circulates,* this is to say something different from where power *resides.* Aspects of power available to human control may have been transferred into cybernetics and networks without a full recognition of the consequences. Communications and cultural studies discourses cannot posit the importance of the "continuous plane" of circulation achieved through the use of electronically mediated communication without also coming to grips with the fact that the "space" of power, ·
metaphoric or concrete, has shifted. In the social imaginary, space used to equal distance. This equation has been "solved" by replacing it with a new one, "space = movement," where movement means continual transit across the old space of distance.[19] In this sense, power is now in continual application and is asserted by movement. Power is no longer most vulnerable when visible, but when it stops moving. Circulating within an electronic grid, only if the lights go off, or the grid fails, is power most vulnerable — to the point of disappearing.

The continuous circulation within on-line environments is a new "center" of power in direct competition with the older materially discrete nodes it was originally intended to link. It is a spatialized nonplace, an infinite corridor for messages that has emerged as an ironic center in and of itself, thereby challenging conventional understandings of how power works and how social relations are organized. VEs are like a contemporary illuminated manuscript made into a theater-in-the-round that visibilizes the immaterial invisibility called cyberspace. They valorize the privileging of transcendence and Idealism and fuel a mix of speculation and desire that somehow we might attain the means to merge with data and information. This longed-for consummation will demand further objectification of power seekers' bodies, which will increasingly be made to adopt fully representational status. If, as research already noted in this work proposes, plural selves will represent their multiple subidentities within interactive VEs, they will also proliferate their own objectification as a multiplicity of iconic representations extending different aspects of self-interest ever outward along several spatial axes at once.

Although immersive technologies offer users an iconographic field that invites a merger with their virtual icons — and with a world of aesthetics and form — these technologies seem more like an anti-aesthetic or anaesthetic. Aesthetics' root in perception connects it to the human body. If I may borrow from Kant to suggest that aesthetics is also about comparison and therefore judging, then judgment requires an embodied dimension. A lived world based on communication as extensibility across space, and a worldview conceived in understanding language only as calling us beyond our physical selves as sites of limitation and restriction, invite an ethical collapse that reduces the space for both subjects and intersubjectivity. Motility, human voice, ethics, and politics are replaced with a set of technical practices and metaphors whose original intentions to bridge the multiple gaps between alienated selves, selves and world, and selves distinct from bodies have been forgotten. A part of the lived world is given authority to stand in for the whole as a result of virtual technologies' reproduction of the process of perceiving the real. As an institutional fact, however, VR does not have meaning attached to it in and of itself. The meanings accruing to it are produced through social relations and therefore remain mutable and subject to critique and renegotiation.

Although to differing degrees credit and money have always been virtual in some sense, Friedberg (1993) links the accelerating shift to a credit economy and its reliance on the virtual buying power of "plastic" to the emergence of VR and invisible "data highways" as the "new frontiers" within this economy. " 'Virtual' has entered the vernacular as the present predictive" (110). In an American Express advertisement from the early 1990s, the AMEX card is portrayed as a "virtual credit space" (Schulz 1993, 437) and is shown holding up a bridge within a natural setting. In suggesting that society is becoming "an identity commodity... trapped by debt" (438), the ads also preview a reverse embodiment, a telematic wraparound of the real world by the visible representations of informational space. Textuality and iconography are rendered akin to a powerful mythic narrative to which living places and bodies must conform according to a pre-scripted set of norms and conventions; in other words, they must accord to something quite similar to information. When people become symbols, any ethical urgency that they be treated as human beings is reduced (Williamson 1978, 167). From the fiber-optic privilege of cyberspace, the human body becomes one more thing within a nihilistic perspective of data overload. The temptation to engage with the new virtuality—even, following cyberpunk and hacking, as forms of resistance—must append a caveat that also calls the imagination back to the embodied physical world. An appropriate phrasing for any message recalling the human body in a virtual age might read, "DON'T LEAVE HOME WITHOUT IT."

Epilogue: Digital Sensations

... to see clearly is poetry, prophesy, and religion all in one ...
John Ruskin, *Modern Painters*

"Is it a virus, a drug, or a religion?"
"What's the difference?"
Neal Stephenson, *Snow Crash*

VR is several things at once: an applied philosophy, a technology, and a socio-spatial practice. It culls from amenable discourses and understandings and both responds to and stimulates long-standing and novel cultural aspirations and desires. Informed by aspects of Platonic and Neoplatonic thought, VR synthesizes an empiricist system of belief with variations of poststructural theories of identity. As such, VR is a "mix and match" technology, adapting various practical and philosophical concepts that in some way are amenable to being built in to the technology itself. Its borrowings, however, are not without implications, and my project in this work has been to discuss some of them.

The current wave of interest in cyberspace and its applications has encouraged promoters to describe it as a new frontier, one open to exploration as well as colonization. To date there is no single paradigmatic virtual environment, though Biocca (1992b, 24) argues that the popular response to what I have identified as the *idea* of VR connects "with some deeper, almost primal desire for freedom from physical constraint, from reality itself." As an applied philosophy made practicable through technology, VR manifests human desires and wishes. Nevertheless, it is not possible to comprehend fully the myriad ways in which virtual technologies and the VEs they permit will affect and be affected by the real world into which they are now being inserted. Conclusions broached through the many related strands of arguments developed in the chapters of this work are best understood as informed by a subjunctive approach to an "as if" technology.

The project of this book results from my long-term interest in both the increasingly pervasive cultural fascination with images and the cultural willingness to readily accept images as equivalent to, or the truth of, what they represent. I have narrowed this focus to examine representational forms in VEs and, to a lesser degree, other nonimmersive ITs. I have been intrigued to understand how images, and therefore truth, might now come to be identified with space itself, and I have asked which broad strands of thought might be woven into an intellectual history that would explain virtual worlds and why they are portrayed as utopias even as they demarcate new spheres of social control. Where is the place of our bodies with respect to the immaterial domain of cyberspace? How do VEs affect or change the Cartesianized relationship between "mind reality" and "body reality" — which together constitute the basis from which we produce meaning? How will access to meaning be altered as a result of the increasing naturalization of the belief that the self seeking greater wholeness and completion might best confirm itself by an imaginative extension across space via these immersive technologies? It seems a strange irony that the search for completion and connection implicates a repudiation of one's embodiment, but as I have argued throughout these pages, VEs are the contemporary version of the immaterial "wherein" one might finally reach after leaving the darkness of the earthly cave and ascending into the light. Subjects borne of "the light within," and bearing a modern duty to produce their own truths, may feel a calling to (as it were) psychically suture themselves to an image technology positioned as the location of truth. Yet as products of a most visual culture, VEs still confirm that "the other" — the object of desire — always remains "over there," just beyond reach. If he, she, or it cannot physically be attained, it is possible to communicate across the distance by means of a device. In time, assuming the desire to join with the other remains strong and unrequited, "we" might even build a machine that seems to conjoin the representations, the messages, the symbols, so that within VEs it comes to seem as if one is standing in the same space with the object of one's desire.

The real and embodied places of the world — and their synthesis of meaning, nature, and social relations — are central to grounding self-identity and organizing sociocultural practices in a coherent fashion. It remains the case that most people on the planet acknowledge the reality of a natural, living world and our meaningful and interdependent engagement with it. At the same time, however, technology helps foster an estrangement from this world, which flows from technology's utility in "pushing back" an often hostile natural world that for millennia was perceived as limiting human intentionality. VR creates a world of spatial representation in which our bodies, always existing along a freighted and leaky continuum between nature and culture, have been set aside. Although cultures organize their members' conceptions of the natural world, each of us, as sentient individuals and in groups, engages with a natural world exterior to ourself. This is how we grasp the meaning of space and how it interrelates with real places. The space outside our bodies and the space of our bodies form a dialectic by which we make meaningful sense of the world. Who and where we think we are depends, at least in part, on how space is conceptualized. If, given the ongoing proliferation and social embrace of electronically mediated communication, individuals increasingly believe that significant components of their identity are capable of relocation "within" communication devices such as on-line ITs and VEs, then the ways in which these people relate to space and their place on this earth will reflect this belief. VR instantiates a representational aspect of this earth created, as it were, to render virtually certain the yearning expressed in "earth designed for 'man.'"

In on-line environments, any "proof" of existence depends on communicated images or symbols of self-identity. "Proof" of our physical dimension is made to depend on an informational subset of existential reality; that is to say, in VR, users communicate the fact of their various existences symbolically across a representational space itself made experientially accessible through iconization. In a double recursivity, motion or information flow understood as communication stands in for the broader existential reality of which this flow is only ever a part, along the way managing to replay the old Platonic parlor trick in which the head swallows its own body. Representation, depiction, and metaphorics replace the natural world, and mind replaces body. The ideal is made more real than the real, with all the "transcendent vistas" this opens to view.

A culture that tries to reduce bodily aspirations, needs, and functional rhythms to spatial models such as those found within the virtual environments of VR is predicated on instrumental rationality and has become a "culture of technique" in which successful and measurable efficiencies have been confused with "the good" (Lovekin 1989). Within a culture of technique, technique becomes an end in and of itself; there is a reliance on the continual busywork of events and their production without due consideration being given to the processes toward which those events, at least in theory, contribute. As a result, means become coterminous with ends, and the techniques themselves become *desirable* standards that are then substituted for meaning.

Place offers a possibility space for grounding ethics. There is a resistant materiality to the natural world; it pushes back on us, and we therefore face material limitations around which we must negotiate, compromise, and move. It is by doing so, however, that such constraints contain the seeds of opportunities. VEs seem to subvert this resistant materiality, and they contribute toward undermining our grasp of it. They suggest that the lived world need not be embraced but simply reprogrammed until it matches "our" desires. As a strategy of technique, however, this sugges-tion — part of the hype of VR — skirts the issue of the desire of the elite "our" who have the ability and power to decide what gets built and how it gets programmed and reprogrammed. The success of the culture of technique, as reflected in its technologies (operating as part of contem-porary everyday life) and the wide array of utilitarian benefits they have provided, allows a blurring between ethical considerations of the lived world and the facts of the "objective" world positioned by instrumental rationality as lying "out there" beyond us. Better, however, to engage the material world in such a way that the imagination is able to conceive of the opportunities that lie in so-called constraints, rather than to accept uncritically any virtual and freighted suggestions that bodies need not matter, that there are few limitations that could not be wished or pro-grammed away, that transcendence comes without much work, or that we need not worry overly about the effects of our actions, since any ill effect can easily be righted by a manipulation of lenses, programming codes, and bandwidth.

Much as contemporary rational "man" might wish otherwise, related yearnings for a spirit world and ritual practices do not appear to surcease, as the hype surrounding VR suggests. In fact, together these yearnings inform the very technologies that seem to be on the cutting edge of tech-nology that science has begotten. It is incumbent on social theorists who study the contemporary intellectual and cultural milieus of late-twentieth-century life to consider an experience of the spirit world as existing for a great number of people, as helping conjure and making sensate a fantasy so powerful that it must be made virtually or experientially real. Imagi-nary transcendence is made more desirable by a rational and empirical system of belief and knowledge organization that denies holism between mind and body. In this system, the mind centers meaning and is partitioned from the subject's body, which is then, like "nature," judged as an artifact and hence potentially worth superseding or "moving beyond." At least since Descartes, this alienated dynamic has operated on an imaginative level and is furthered by a belief that understanding primarily consists in forming and using appropriate symbols as representations. VEs expand the purview of this dynamic by suggesting that surpassing bodily limits might incorporate a spatiality somehow existing on "a separate plane."

A metaphysical answer to the question "To whom do we now communicate via metaphorics if not to God?" is suggested in the artificial intelligences described in William Gibson's science fiction discussed in chapter 1. The aleph/null of the cybernetic god that Gibson describes, and the widespread fascination with his speculative fiction within the virtual research community and a wider mediated public, suggests that the desire for an Absolute and the transcendent plane continues to inveigle significant aspects of Western elites' imaginations. The synthesis of Platonic and Neoplatonic *points of view* within the computational realm is the vision and sight of the cyborg operating as the "impure" hybrid of nature and culture achieved by fusing material technology and symbolic language. In all of this, the underlying text-dependent mechanisms of cyberspace are connected to the visual display of a VE in a fashion similar to the Hellenistic metaphors of light authoring belief in, and construction of, the spatial relationships that they had also first imagined.

This spirit and metaphysics more generally are never fully permitted or acknowledged to exist separately from social relations, yet the interests of capital, for example, are very much present in the rush to assuage the ailing psyche of the contemporary virtual pioneer. VR technologies allow that part of the Western social imaginary that connects a wish for extensibility beyond the coordinates of our bodies and fantasy to achieve a measure of psychic balm. Fantasy that addresses those parts of the psyche that are in "excess" of that required to fulfill the demand to inhabit a coherent identity can be allowed as a kind of personal, interior truth. This truth's manifestation or production becomes morally permissible within the virtual world that promises an ability to consume — if only images — that virtually fulfills capital's promise and premise.

On a related note, Fiske writes that "middle-class aesthetic/critical distance is a homology for the distancing of the bourgeoisie from the labor of production" (1991, 97). In this regard, virtual technologies have the potential to cut both ways. In offering an aestheticized representation for capital removed from its production, these technologies have the strong potential to reify the distance between commodities and the laboring bodies who are their makers. At the same time, by their superior verisimilitude and speed, and to the degree that these ironically disembodied representations model surplus and value and make visible the movement of "surplus value," virtual technologies are able to accrue power to themselves and suggest they are a site of production. This is a carnivalesque hat trick on Marx, a "sight gag" that inverts base and superstructure by suggesting that culture is base. Following this, the nature of the materiality on which culture depends — in this case the data bits of information in ITs and electronically mediated communication — is more easily naturalized and thereby set aside as a subject of critique. Information holds a unique position under the "sign" of economy. Clearly there is a distinction, for example, between a two-by-four piece of lumber transformed by labor and the tree from which it was milled; nevertheless both the two-by-four and the tree are material objects. This is arguably not the case for information, the nature of which is not fully amenable to materialist explanations, even though there are obvious material consequences to its circulation. Information is as much a cultural production — perhaps more so — as it is a material reality, and its reproducibility, ephemerality, and transience, combined with its increasing economic centrality, suggest a need to rethink the place of ITs within a globalized world economy. Stated otherwise, what seems to be a "pure" conduit for the communicatory "flow" of information has itself become a production site.

In 1916, John Dewey could already write that "society not only continues to exist *by* transmission, *by* communication, but it may fairly be said to exist *in* transmission, *in* communication" (5), an argument connected to his belief that thought is possible only in the presence of language. Virtual technologies now suggest vision's technical relocation to *within* the *technologies* of transmission. We may therefore expect the digital world of communications to increasingly reflect a technical elite's accelerating impatience with the physical body "parked" on the other side of the interface. Already, this elite increasingly argues the rational efficiency of a further minimization of bodily motility and touch through which people in part make ethical and political sense of the world. As discursive practices, VEs exemplify links between such "efficiencies" and contemporary theory based on unacknowledged desires to confuse bodiliness with *ideas* of the body, and our bodies' movements with *ideas* of bodily movement. The rush to fuse representation (and culture) with the broader lived world—to "write" the former over the latter so as to suggest the "death of nature" and the "triumph of the cultural or the text"—itself depends on a use of communications *technology* that remains undertheorized. Future generations are likely to be "stupefied" at the "postmodern orthodoxy" that human bodies are primarily "if not entirely" linguistic and discursive constructions (Hayles 1992, 147). Fantasies of disembodiment are socially diffused more so today than at any time since the Middle Ages and are intimately linked to contemporary technologies (Hayles 1993a, 173). The orthodoxy referred to here is captured in the notion that "the body is a (social) text," and, at least in part, in the creative contradictions of modernity noted by Latour and set forth in figure 3. A belief that society is our free construction, coupled with the assertion that human bodies are largely socially or discursively determined, leads to the second "guarantee" Latour identifies: even though we do not really construct society—the human body included—we act as if we do. We can only maintain this belief, however, by also holding apart the spheres of nature and society or culture. Therefore human bodies are conceptualized as fully apart from nature. The almost explicit schizophrenic hubris here is counterposed against the dynamic instituted under the first guarantee suggested by Latour: that is, we mask our agency in constructing nature by acting as if we do not control it. VEs are part of this fragmenting masking. By offering a programmable depiction of nature, our bodies included, they illustrate and become metaphors for the recursive myth that "nature is as if we do not control it." This dismissal or reduction of actual human bodies, which are either written off as category mistakes, lost in the gap between these modern dichotomies, or made suzerain to their own representations crafted by alienated, conceptually disincorporating subjectivities, is also a dismissal of "bodies politic."

Understanding our bodies as existing along a leaky frontier between nature and culture suggests the limitations to theorizing completely hard-bounded distinctions between people and places. This makes the human body a messy subject for theory, yet one worth pursuing. Our bodies are rarely inert in the static sense of "the body." If I argue for my body as a place, and both my body as a place and the place within which I find my embodied self spatio-temporally to coexist within "nature" and "culture," then there will be bodies and places outside of history, within history's organizational purview, and in both positions at once. It is far easier to theorize a space—such as a VE—that has been thoroughly historicized by notions such as coordinated mapping, absolute space, relativity, pleasure, and so forth. But when we marginalize human bodies, which fit awkwardly, if at all, into spatialized theories and VEs, we reduce situated, intersubjective individuals and communities to measurable sets of spatial coordinates that are always open to the dangers of expropriation, and we contribute to the imperilment of both nature and culture, wherever the "frontier" between them may be drawn.

Feeling uneasy with the larger world situated on the other side of Elias's modern wall, an increasingly privatized and fearful subjectivity reconceptualizes the world as information, thereby justifying to itself its disengaged retreat into the so-called "cocoon"—the insect metaphor so complementary of hive mind. Finding itself in a vacuum, the larval subject produces new forms of spatialities, imaginary architectures, cosmographic mappings, and fantastic digital landscapes more amenable to its self-perceived sense of strangeness and disarticulation (see Vidler 1992). A "psychic homelessness" informs VEs as the modern carnival turned grotesque—a disconnected, perhaps monstrous, but also recursive, language game in which machined laughter mocks the environmental evidence of tangible bodies. The focus on VR as a polyvalent world of pleasure notwithstanding, the need for limits, standards, and conventions (required to make any form of communication, technology, and certainly communication technology operate in a meaningful way, whether its use is pleasure or otherwise) will condition the spectacle of cyberspace equally as, for example, the editorial decisions forced by the forms books take. If anything is implicit in the present work, it is that ideologies or developmental logics that underlie the intention behind technologies then get built into the technologies themselves. The limitations of each technical form constrain and empower what it does. Here lies a difficulty in a shift from metaphoric understanding based on narrative to one based on images and living in visions. VEs set aside the importance of a temporal dimension implied in narrative and implement a programmed illusion of a potentially infinite, spatialized present. If Gibson's science fiction novel *Neuromancer* gave voice to a virtual community, it did so in a print format arguably closer to orality than the VR "picture writing," which is part of its fascination for *readers,* and for those who then listen to the stories readers tell about the yarns that Gibson spins.

It is worth holding on to the connection between voice and orality and by implication that people may come into a fuller knowledge of things when they physically speak to each other about them. "Giving voice to" admits that it takes time for someone to first have spoken, others to listen, later to rebut, reply, and circulate. To give voice implies a discursive community different from the politically neutralizing isolation that telematics, however extensive of the self, may well imply.

Yet to date, resistance to unwanted political use of technologies is often predicated on an implicitly unitary individual situated at a remove from the technology under review. Although I am arguing for identity in this work, I do not suggest that there is a unitary identity. Unitary identity resonates with the iron cage of logic and is to be resisted in equal measure to suggestions that identity is outmoded conceptually and life is largely a series of performances through which we negotiate and network. The author/actor intrasubjectivity seems more amenable to an engaged citizenry and civic practices. Further, it is worth noting that the kinds of polyvalent pleasures available in VR speak to a false distinction between identity and unconscious—the unconscious (like the psyche) here implicitly positioned as all those parts of the self that are in excess of what dominant cultural conventions allow an identity to be. Positioned thusly, a so-called unitary identity could only ever be partial and always is threatening to come apart. In such a context, VR is a machine for transcendence that delivers the overly atomized modern individual into a merger of subjectivity and landscape, a scene in which he or she potentially may communicate with other disembodied selves gathered in a virtual room but situated anywhere around the globe. These possibilities challenge traditional notions such as Gramsci's organic intellectual who finds explanation and meaning within his or her specific cultural background (Cavalcanti and Piccone 1975, 55) with respect to how resistance to a technology's undesirable political implications might be constructed. So too—because of the requirement to actively don or engage with the technology physically—does the rendering complicit of users' acceptance of the power dynamics underlying VR call into question not only the political dichotomy of ally and opposition but also that of pleasure and surveillance.

A lack of sufficient recognition of the dialectic within which pleasure and surveillance can operate within electronically mediated communication runs alongside the academic emphasis on pleasure, which I would suggest has reached a point of saturation, and even overaestheticization of the concept. For many theorists, this precludes in advance considering the political implications of optical and information technologies beyond the pleasure principle and the "politics of pleasure," which is too often also the individuating politics of "me" or even the anti-politics of "I like to watch." Abandonment of an ethics of scale is central to this lack of recognition. When individual or bodily pleasure is emphasized, patterns of surveillance across systems often remain underconsidered. More work remains to be done here, but if scale is brought "into the picture," a strong tool and cultural technology can be brought to bear on understanding how surveillance capacities and pleasurable effects can appear perfectly in keeping with one another, ironically or otherwise. For example, recalling earlier discussion of the instability of the meaning of visual images, it is, arguably, this instability (read polyvalency) of the image that, in part, attracts individuals to VR's complex promise. This instability or overdeterminacy allows the individual user a kind of control — to designate in advance, through the use of language, how he or she will be depicted during the VR experience while not requiring any coming to terms with the finite frame that circumscribes a VE's polyvalency. In this way, VR acts in a similar fashion to a Marcel Duchamp "readymade": a writing or graphing of technology onto culture so as to lessen the sense of loss of individual space so palpably captured in Elias's spatial metaphor of the "wall" erected by modern individuals between themselves and the world. Further, the icon of a body part — the virtual hand inserted into the user's field of vision — confirms the user's existence, yet this confirmation communicates visually across the interface. This interface is not only where surveillance is operationalized; it is also a gap and thereby always potentially as nostalgic or as suggestive of loss as it is suggestive of a future completed by technology.

If one looks for pleasure, one will find it, voyeurism included. So too for surveillance; however, focusing only on pleasure minimizes recognition that those being "pleasured" are also under review, their actions capable of being recorded as patterns of data available for analysis and forming the basis for taking action on the part of those watching and recording. This is precisely the intention of the Army Research Lab's simulation reviewed in the introduction. The individual soldier-trainee, operating within the virtual killing theater of the display, receives "pleasure" and reward from each enemy annihilated and each helicopter shot down. At the same time, however, the program records her or his performance, reactions, and decisions taken as part of military discipline and control.

The politics of pleasure notwithstanding, while virtual social organization might take on a more collective ability, not unlike some of the more progressive understandings advanced by the cyberpunk counterculture, this *form* of resistance would depend on mediated standards and hence a new form of centralized power — the rhizomelike conduits of information that under the auspices of "world economy" become the new sites of production and power. A strong sense of decentralization has been ascribed to the rhizome: its horizontal organization seen as inherently more democratic and overdetermined than older forms of vertical integration. The logics of late capitalism, however, variously work to produce strongly centralized control *through* decentralizing tactics, the use of ITs included (Harvey 1989). What the rhizome metaphor fails to take into account is that ITs permit dispersal precisely because they also facilitate monitoring of all the lines of communication (see chapter 1, note 2). Like the panopticon's guard, who may or may not be watching the inmates, such monitoring need not necessarily be occurring at any one moment or place; nevertheless, the fractured flesh of spatially isolated citizen/consumers dependent on ITs seems less than benign. Decentralization as contributing toward centralized power suggests an ambiguous eventuality that might demand a painful and lengthy reconsideration of the modern distinctions erected between humans and their technology, even between existence and communication. Are these distinctions worth retaining within what, politically, might become an extended struggle over codes and the desirability of becoming cyborgs? I believe they are, but I understand that many industry and military players and academic theorists do not. Do "we" value the space of individual subjectivity, or has the weight of responsibility placed upon it become such that certain individuals would gladly give it over to an electronic hive mind based on the "truth" of propositional logic — one "wherein" the "correspondent" representation is increasingly more real than the real?

Finally, just *who* is this "we"? Who is so eager to cede subjectivity and the struggle to gain identity to the hive mind planetary soul ITs make accessible and the string of addictive repeat performances, or workerdrone return flights to the hive this implies? An illegal Latina maid working under the table in Westside L.A.? A Colombian lesbian living illegally in the United States who dares not claim refugee status lest she be deported back to Cali death squads? A newly emergent black political class in formerly apartheid South Africa? Somehow, I think not. Most of the world still struggles for the means to practice the embodied subjectivity that Western technical and certain academic elites would now discard as an outmoded Enlightenment commodity. Despite the promise of an infinite potential of electronically generated digital patterns, pixelations, and performances, the "as if" geography of the virtual world is not really "whatever" we want it to be. To date, it is rather small, both materially and experientially.

The promise and hype of VR and ITs more generally is part of an ideology of the future, produced in an amnesia and loss of history that forgets the broken promises of past technologies such as the "universal educator" (TV) and "too cheap to meter" (nuclear power). Metaphors of progress and evolution work to suggest that bodies and places are always incomplete, partial, and by *necessity* thereby flawed. Perfect vision and absolute clarity lie always in the glittering future, and the move to all things virtual, positioned as a solution to the "problem of the present," or the "problem of the body," updates Horace Greely's admonition, linking movement through space to progress and the future, "Go West, young man, and grow up with the country." But if understanding can always only be partial, and if the mind is also flesh, then answers cannot lie solely within the transcendent light and reflected images inside VR's head-mounted display. Optical technologies may allow us to "see through a glass, darkly," but as Paul reminds us, "face to face: now I see in part." Although they remain a minority, those advocating relocation to an electronic "frontier" would abandon the wider world, and like a circled wagon train, face on-line facsimiles of themselves signaling one another that their cartoonlike representation of the social and natural world is finally complete.

Notes

Introduction

1. The term originates in Husserl's distinction between the scientific world and the *Lebenswelt,* or "lived world" (Reese 1980, 239). In English, *Lebenswelt* is often translated as "lifeworld," or at times the "taken-for-granted world." Whichever term is employed, each directs "attention to the significance and entirety of a person's direct involvement with the places and environments experienced in ordinary life" (*Dictionary of Human Geography* 1994, 331). I understand "lived world" to refer broadly not only to those natural, socioeconomic, and ethico-political environments that constitute the continuum running from private to public spheres, or from nature to culture (virtual technologies included), but also to our selves as *existing* embodied individuals and as people in groups.

2. Internet Relay Chat (IRC) is a "place" "where anyone with Net access can go to for free, real-time chat, 24 hours a day, year round" (Quittner 1995, 119). It is modeled on a series of virtual text-rooms organized by topic and issues, and "in" or "on" which geographically disparate individuals conduct on-line real-time discussions.

3. David Porush (1996) suggests that utopian ideals contain an implicit design offering a "feedback controlling machine," which he explains in the following manner: "As technology manipulates and alters human nature, and human nature adapts itself to the new technosphere, new versions of utopia arise, which in turn promote new technologies, which in turn change the context for defining human nature, and so on" (122). Porush's understanding is similar to David Rothenberg's (1993) concept of "the circle of technology" — an interpenetrating dialectic among human intention, new technologies, and the new intentions they help authorize.

4. The conflation between the promise embedded in "just around the corner" technologies that in reality are far from extant and those that are already

"up and running" is not restricted to VR. Visual artist Paul Couillard, for example, working at the Holography Museum in Toronto, found that gallery visitors often expressed disappointment with the examples of advanced holography on view. As a consequence of watching *Star Trek: The Next Generation,* many of these relatively sophisticated individuals believed some variation of the starship *Enterprise* holodeck to already exist (personal communication with artist).

5. To gain personal experience with VEs, I demonstrated immersive VR technology at the *Spring VRWORLD '95 Conference and Exhibition* at San Jose, California, 22–25 May 1995. Working for Division Incorporated, I demonstrated a variety of interactive and immersive VEs: the Weapon Systems in Virtual Trials application described in this chapter, a British Ford automotive showroom, Matsushita Corporation's virtual kitchen and virtual house, the Gulfstream V corporate jetliner passenger cabin, an offshore (North Sea) multilevel oil-drilling platform, a virtual repair facility for McDonnell Douglas F/A-18 jet engines, as well as a VE that permitted "walking through" the human immunodeficiency virus (HIV).

6. It is possible this trend may slowly change if VR games follow the course set by PC gaming. In 1997, according to PC Data Inc., seven of the twenty top-selling PC computer games among all age groups were "kill-or-be-killed" titles. Among preteens, however, games featuring the Mattel doll Barbie, Barbie Fashion Designer and Barbie Magic Hair Styler, topped the best-seller list. Increasingly, software designers are creating games that are "low on violence and high on relationships" (Lohr 1998, D1), and Broderbund's Carmen Sandiego series, in which children play trivia games based on geographical and historical knowledge, is a perennial seller. Nevertheless, as the title of Trendmaster's War Planets: Age of Chaos indicates, violence and killing the enemy continue to be an important factor in current PC gaming.

7. A separate personal immersive VR experience — provided here as a point of comparison — took place at the Club Mendota, a large bar on the west side of Madison, Wisconsin, located in the rear parking lot of the toniest mall in town. A gang of young guys is waiting to try their hand at playing Dactyl Nightmare — a VR game promoted in clubs like this around North America. I arrive with a friend who has agreed to try the technology with me and discuss his experiences afterward. We're the only people over thirty. There's one woman with her boyfriend, and they're with two of their male friends. It costs five dollars a play, three for students during the happy hour extending from 4 to 8 P.M. We sign a sheet and wait our turn. There are about ten people ahead of us, so we stand in a semicircle with the others and examine how the game is set up. It consists of two pods in which the virtual contestants stand. They're ringed all round with a waist-high padded railing that I will discover is useful to prevent oneself from

falling over. The audience can participate somewhat in the players' experience via twin side-by-side high-resolution color monitors that replicate the players' points of view, and what they see within the virtual field. When our names are finally called, we step up into the opposite pods. I first put on the position tracker, a device about half the size of a portable CD or tape-cassette player, which is attached by cinching a belt around my waist. I then don the HMD, which is cinched very tightly around my head, both downward and inward. I'm holding a "gun" or joystick device. If I press the button on the front of the handle, I can "fly" forward in the direction I am looking. Pushing the button on top of the handle fires the gun that releases a virtual charge designed to eliminate either the "opponent" or the pterodactyls that circle overhead from time to time in search of prey.

The game lasts three minutes. If I look carefully into the "eye phones" of the HMD, I can see a digital readout that tells me how much time remains. But it's hard to see, and I don't bother — the resolution being much less sophisticated than the military and major corporate and medical applications on display at industry trade fairs. At first, the sense of vision seems restricted. It's like looking down a short tunnel into the stereoscopic TV screens. In the bottom left corner of my visual field is a cyberspace hand, but it's not really connected to my own — again, unlike more sophisticated applications where my virtual hand in virtual space responds in kind to the movements of my real hand in real space.

The world the game creates is a kind of meta-chessboard in space. In the middle of a central platform stands an Acropolis-like structure minus its roof. Elsewhere in this space are freestanding Grecian columns positioned at random intervals. They are shaded and can be negotiated with relative ease. At the center point of each side of the square or "central space" is a stairway, all four of which rise to equally sized platforms smaller than the central one below. The floors of all these spaces are set out in a geometric tile pattern. Green conical shapes give the impression of cartoon trees. These abstract shapes are premised on the fact that the mind will accept an image as "virtually" real even if only traces of a sense of reality are present. What is more disturbing here, however, is that the game is a virtual killing ground.

During the three minutes of play, I have difficulty remembering which of the buttons does what, so I'm constantly firing off bullets when what I mean to be doing is moving forward in order to experience the novel sense of motion. The program circumscribes certain actions. While waiting to play, we notice one person turn the gun around and attempt to shoot herself, but the game will not allow this. The software momentarily pauses, then repositions the image of the player straight ahead. During play, at one point I am on one of the upper platforms and try to jump off, out through "space," and down to the platform below, but the program won't let me. I have to turn my head rightward, toward the stairway and fly/move over to the top of it, then aim the button in the down-ward (straight ahead) direction. I spend most of my time looking around the VE. My friend later tells me that he spent most of his time trying to locate my image within the VE. There is a lot of noise, New Age sounds, and twice I'm captured by the pterodactyl and carried skyward. The bird carries me/my point of view high in the air and then drops me to the ground. The image I see comes to a shuddering halt, and I am eliminated. In carrying my point of view skyward, the program also positions an image of myself in the virtual space as part of what I see, so that I follow or fall to earth just behind the image of my virtual body as it drops down to the platform and shatters. Yet immediately afterward I am back on my feet again, like the cat with nine lives.

8. The term "Fakespace" is already in use. The Fakespace Simulation System, or FS2, is a stereoscopic immersive display, a variation on the HMD.

9. This is a key point in David Harvey's (1989) argument about post-modernity.

10. See, for example, Benedikt 1992b; Biocca and Lanier 1992; Laurel 1993; Stenger 1992.

11. Thomas Hobbes's distinction between the Author and Actor is central to his definition of "Persons Artificiall." A "Naturall Person" is one "whose words or actions are considered, either as his own, or as representing the words or ac-tions of an other man, or of any other thing to whom they are attributed, whether Truly or by Fiction" ([1651] 1985 217). Some "Artificiall Persons," however, "have their words and actions *Owned* by those whom they represent" (218), and these individuals are Actors. Those who own the words and actions are Authors. Actors act on the *authority* of the Author, and should any law of nature be broken by the Actor on the command of the Author, then, according to Hobbes, it is not the Actor but the Author who "breaketh the Law of Nature" (219).

1. A Critical History of Virtual Reality

1. Kevin Kelly, promoting his vision of society modeled on the collective intelligence of the beehive—the *HiveMind*—writes that "a recurring vision swirls in the shared mind of the Net, a vision that nearly every member glimpses, if only momentarily: of wiring human and artificial minds into one planetary soul" (1994, 24). In passing I note the similarity to Emile Durkheim's *conscience collective,* defined in *The Division of Labour* as "a set of beliefs and sentiments common to the average members of a single society [that] forms a determinate system that has its own life" (in Lukes 1972, 4). Like the emerging forms of ITs, networked communications systems, and VEs, the *conscience col-lective* is only realized through individuals yet is distinct from individual con-

science. It inheres to a "psychic type of society," is "diffused throughout the whole" of that society, is "independent of the particular conditions in which individuals are placed," and "results from fusion of individual impressions" (4). Durkheim's critics disliked the metaphysical nature of the concept (see Gane 1988), pointing out that it blurred distinctions between the moral, the religious and the cognitive. In this it anticipates the "electronic sublime" (Carey and Quirk 1969–1970) which telematics achieves in its marriage of computation to telephony. The networked conflation of morality, religion, and cognition parallels the collapse between culture and information, and between culture as a commodity and a form of life that thereby resists the reduction implied by commodity. The conflation suggests the need to rethink these categories.

2. At the outset of this work I named VR as a machine to realize desires for transcendence. Deleuze and Guattari (1987) and Raulet (1991) write of "desiring machines" — at the very least as a subset of a "collective assemblage of enunciation, a machinic assemblage of desire" (Deleuze and Guattari 1987, 23) that would allow us to "arrive at the magic formula we all seek — PLURALISM = MONISM — via the dualisms that are the enemy, an entirely necessary enemy, the furniture we are forever rearranging" (20–21). These passages echo Durkheim's *conscience collective* discussed in note 1 to this chapter. They also well describe the merger with a collectivity the modern Western(ized) self may seek within VR. Access is individuated. "We" are all together online, yet home alone via the "dualism" of binary logic, the "furniture" of mathematical codes which permit the constant "rearranging" of the so-called picture language within VR's representational and emblematic "space."

Yet a "desiring machine" already has swallowed the subject that Deleuze and Guattari also identify as representationally coeval with the state. Even an intermediate stage of political agency such as the man-machine cyborg seems unavailable to their approach for those who would choose not to cede subjectivity to the machine at this historical juncture, given the appetite implicit in "desiring machines." Raulet grasps that desiring machines efface locality within a seamless web of *network* — the rhizomal structure Deleuze and Guattari privilege. Desiring machines, ironically, are an anthropomorphosis that occludes humanity, let alone a reconsideration of the political complicity of the subject re the state. Though rhizomes are an ideal metaphor for the content/form of modern IT and telematics, rhizomes-as-metaphor reproduces the power of representation Deleuze and Guattari seek to undermine.

Although representative forms are essential to communication, their excessive use is worth resisting; and VR's current developmental trajectory manifests many aspects of such excess. Machines *for* transcendence intends to skirt the often in-

advert metaphysics that attends analysis of these issues by suggesting a relationship between human agents and technology equally as I acknowledge that
technology now inflects human *existence*.

3. However crude early flight simulators may have been, they nevertheless
were intriguing enough to merit outside study. Randy Pausch et al. (1992) note
studies dating from 1939.

4. Margaret Morse (1994, 160) makes a similar observation. "While the
process of identification associated with the cinema paradoxically depends on
distance... [immersive technologies] involve introjecting or surrounding the
other (or being introjected or surrounded) and ultimately, the mixing of two
'bodies' in a dialectic of inside and outside that also can involve a massive difference in scale."

5. Anticipating both Gelernter's "Mirror Worlds" and Sutherland's "Ultimate Display" by more than two millennia, Plato writes:

> "Just a minute, and you'll be more surprised still. For this same
> craftsman can not only make artificial objects, but also create all
> plants and animals, himself included, and, in addition, earth and
> sky and gods, the heavenly bodies and everything in the under-
> world."
>
> "An astonishing exhibition of skill!" he exclaimed.
>
> "You don't believe me?" I asked. "Tell me, do you think that a
> craftsman of this sort couldn't exist, or (in one sense or another)
> create all these things? Do you know that there's a sense in which
> you could create them yourself?
>
> "It's not difficult, and can be done in various ways quite quickly.
> The quickest way is to take a mirror and turn it round in all direc-
> tions; before long you will create sun and stars and earth, yourself
> and all other animals and plants, and furniture and the other objects
> we mentioned just now."
>
> "Yes, but they would only be reflections," he said, "not real things."
>
> "Quite right," I replied. (Plato 1987, 10.596c–e)

6. Turing asserted that "at the end of the century the use of words and general educated opinion will have altered so much that one will be able to speak
of machines thinking without expecting to be contradicted" (1950, 442). A machine that imitates human intelligence may provide little or no useful service
(Bolter 1984); there are already plenty of humans to do human tasks. Instead
(shades of Marx) the Turing Machine has come to be seen by subsequent inventors and assorted "technotopians" as a kind of defining technology or metaphor
of the age, one that reorganizes the way humankind relates to nature. Humans

become "information processors," nature "information to be processed." Within the "world" of the Turing Machine, computation is "nothing more than *to replace discrete symbols one at a time according to a finite set of rules*" (Bolter 1984, 47) (synopsis drawn, in part, from Bolter 1984, 10–14, 43–47).

7. Nelson 1977, 120–23, as quoted in Rheingold 1991, 91.

8. Information about the GreenSpace Project was obtained, in 1997, from the project's Web site at http://www.hitl.washington.edu/projects/greenspace. I also attended a presentation by Jim Elias of US WEST offered in the Televirtuality Session at the spring VRWorld'95 Conference in San Jose, California, 24 May 1995. The session was titled "The GreenSpace: An American Perspective." Robert Markley (1996b, 6) identifies something of the ritual quality that is present in virtual interactions, and how this ritual is already being linked to as-of-yet unarticulated patterns of consumption. He finds the content of GreenSpace is long-distance communication, which the project "subsumes and recodes. The conference itself becomes the product to be disseminated rather than a means to an end."

9. For a chilling account of these applications, see Sterling 1993.

10. *The Perception of the Visual World*, 1950; *The Senses Considered as Perceptual Systems*, 1966.

11. Representations of individuals and their body parts in cyberspace are sometimes called *puppets*.

12. With personnel movement from NASA to educational institutions, this may now begin to change. Other opportunities, particularly at universities, are increasingly available for experiencing advanced VE design, as Hayles's testimony suggests: "From my experience with the virtual reality simulations at the Human Interface Technology Laboratory and elsewhere, I can attest to the disorienting, exhilarating effect of feeling that subjectivity is dispersed throughout the cybernetic circuit" (1993b, 72).

13. Politicized social relations render NASA keen to construct and preserve legitimacy in the taxpaying public's eye. More so than in the case of the U.S. Air Force, there are regular congressional movements to dismantle, downsize, or restructure the agency. It behooves NASA to air its successes widely, given concerns with industrial espionage and national security.

14. Transputers are "very powerful microprocessors (chips) that not only compute, they also communicate with other transputers or processors. They... can be assembled and used like Lego" (Sherman and Judkins 1993, 46). This Lego-like quality permits rapid division of workload within the computer as well as easily added capacity for applications requiring greater processing power.

15. The holodeck first calls on the *idea* of a hologram that already exists within culture, then reworks and extends this existing understanding. Viewers

watching *Star Trek* understand that holography already exists as a scientific invention and is in use in a variety of ways. They seem willing, however, to accept that the holodeck in an admittedly more rudimentary version must already exist (see introduction, note 4). It would seem that if individuals can absorb an idea of a technology into their imagination, they can also come to believe it must already exist. Stated otherwise and with reference to holography, people believe that the technology projects images out into the air rather than onto a surface, as is the case with all forms of photography. Conceptually, the holograph would seem to be magically understood as a projection of pure light that can take on an image without the need for a surface.

16. Within this earlier formula, traditional SF did manage to comment on social relations. The following passage from Robert Heinlein's *Beyond This Horizon* (1942) — a description of a computer — is strikingly descriptive of today's global cyberspatial data flows: "The manifold constituted a dynamic abstracted structural picture of the economic flow of a hemisphere" (cited in Kurland 1984, 200).

17. The issue of fragmented anomie Stone identifies here is further addressed in chapters 4 and 5, which are organized respectively around vision and language. It seems a far cry from the hopefulness and liberatory potential of an Enlightenment sensibility based on vision and its metaphors to the contemporary alienated and exurban landscape where ITs and VEs are written and built. "Shedding light" on the interplay between current technical and social changes is possible by examining the changing history of how vision has been conceptualized from the classical Atomists, through the Enlightenment, to the present.

18. This "community" corresponds to an alternative definition of cyberspace offered by Heim (1993, 32): "The broad electronic net in which virtual realities are spun."

19. No page numbers are provided for Manovich 1995a and 1995b. Respectively, these articles were downloaded in March 1998 from the following Web sites: http://jupiter.ucsd.edu/~manovich/text/digital_nature.html and http://jupiter.ucsd.edu/~manovich/text/Checkpoint.html.

20. In "Visual Pleasure and Narrative Cinema," Laura Mulvey invokes the Freudian "scopophilic" pleasure of looking and being looked at to locate her argument that a fascination with film and the visual is reinforced by preexisting fascinations already at work within the individual subject (1975, 6). Although she thereby acknowledges a certain historic specificity, it is restricted to the individual level by her psychoanalytic take. Virtual and other "psychotechnologies" also trade at this level. Mulvey relies on a Freudian conception of narcissism as fascination with the human form wherein (self) identity ironically is

located in an act of self-recognition with a corresponding image. VEs and ITs promote self-extension. All communication promotes this within an understanding of what it is to be human. However, following McLuhan (1964, 51) — who argued that the West's cultural bias is evident in its misinterpretation of the Narcissus myth as meaning only an injunction against a false self-love achieved through reflection and image — I want to note the link between *narcotic* and *narcissism*, and the numbness that results from an unwise overextension of the self into exteriorized image, such as body-as-information. In identifying self-interest with the screen, the cinema is also an anodyne for an overtaxed subjectivity perhaps too closely identified with reproducing the demanded stability that is a precondition of the state's existence (see Deleuze and Guattari 1987). In a postnational culture, such stability seems increasingly less central to global selves and undesirable to global corporations seeking "flexibility" in their structures and employees.

21. Citation taken from Virilio 1994.

22. From personal notes taken at a lecture given by Anthony Giddens, University of Wisconsin–Madison, 7 March 1994.

2. Precursive Cultural and Material Technologies Informing Contemporary Virtual Reality

1. Haraway's observations were given in response to a question I posed, inquiring how acknowledging technology's agency — intended or otherwise — might be incorporated within contemporary theory while avoiding (1) the accusation and the pitfalls of technological determinism per se, and (2) contributing inadvertently to a ceding of control by humans to machines. The citation is from personal notes. From the AAG 1995 Chicago annual meeting, "Harvey and Haraway: Debate and Discussion," Saturday, 18 March 1995. Notes from the meeting are published as "Nature, Politics, and Possibilities: A Debate and Discussion with David Harvey and Donna Haraway," *Society and Space* 13, no.5 (1995).

2. I would note the Renaissance invention of the camera obscura; the period's rediscovery of Ptolemaic perspective; the codification and application of perspective techniques by Alberti, da Vinci, and others; the development of cartographic mapping and the landscape idea, which depend on enframement and visual techniques to extend the spatial power of the user/subject; and, over time, magic lanterns, the camera, cinema, television, and video as precursors of newer fiber-optic technologies that further extend the power of the eye.

3. The camera obscura — literally, a "dark chamber" — is "an instrument consisting of a darkened chamber or box, into which light is admitted through a double convex lens, forming an image of external objects on a surface of paper,

glass, etc., placed at the focus of the lens" (*Oxford English Dictionary,* s.v. "camera obscura").

3. The Sensation of Ritual Space

1. This permitted the establishment of, for example, the futures market at the Chicago Board of Trade (Cronon 1991, 122, 332), which played a vanguard role in shifting speculative activity away from space toward time (Carey 1983, 316). Cronon (1991, 120) observes that Chicago's Board of Trade was founded in 1848, the same year the telegraph reached that city. Cronon also finds that the telegraph's spread across the United States led to an amalgamation of earlier discrete regional economies. A newly emergent "market geography" was independent of local climate or soil fertility. It relied instead on price and a flow of information throughout the entirety of its wired economic sphere (121).

2. This dynamic might also be understood as the attempt to merge two different forms of knowledge, to weld through the use of telematics "acquaintance with" and "knowledge about," or, stated otherwise, to weld direct experience with mediated learning.

3. Some forms of intimacy, however, would seem more easily achieved without the visual or physical presence of the other person. The example comes to mind of two teenagers who would rather talk on the phone for two hours than actually walk a block to each other's houses to hang out. Before telephony, the late-Victorian exchange of postcards between proximate urban neighbors is another case in point.

4. The pre-telephony place-bound public speech act passes through the intimate space between two speakers. At the same time, this speech act itself can form part of the informational content of other future speech acts that may occur in a variety of places. In contradistinction to the "from here to there" aspect of transmission through space, communication as ritual contains a richer understanding that past, present, and future actions form part of an ongoing process.

5. Krueger, promoting a virtual world, addresses the importance of perception to imaginative engagement in VEs: "Imagine that the computer could completely control your perception and monitor your response to that perception. Then, it could make any possible experience available to you. In fact, it could provide any *imaginable* experience.

"In a sense, an artificial reality is the incarnation of imagination: a projection hallucinogen that can be shared by any number of people. It is a laboratory for philosophy where we can ask basic questions such as, 'What is reality?' 'What is perception?' 'Who am I?' in fundamentally new ways.

"Contemplate the limits of your own imagination. What can you dream that you cannot experience? Think now — or the answer will be provided as a *fait accompli.*" (1991b, xvi–xvii)

6. Aristotle, *De caelo,* 3.2.300b. As cited in Jammer 1969, 11.

7. To wit, asks Lefebvre (1992, 1), is Aristotelian space an empirical tool, or is it superior to sense evidence?

8. Jammer (1969, 9) notes that for sixth-century-B.C. Greek Pythagoreans, space limited or separated different bodies. The fifth-century-B.C. Atomists Leucippus and Democritus held that space was complementary to, and bounded by, matter, and that the two were mutually exclusive (11). Fifth-century-B.C. Pythagoreans — as further discussed in chapter 4 — came to identify the spatial concept of the Void with air itself, thus introducing a rudimentary sense of abstraction and extension into what was meant by space. Jammer (9) cites J. Burnet, *Early Greek Philosophy* (London: Condon, 1914), 51: "The Pythagoreans, or some of them, certainly identified 'air' with the void. This is the beginning, but no more than the beginning, of the conception of abstract space or extension."

The word *kenon* — ancient Greek for "the empty" — was often used synonymously with the word "space" (Jammer 1969, 11). It was a "second-generation" Pythagorean and friend of Plato, Archytas (Reese 1980, 26), who separated Pythagoreanism's number theory from its mystical and religious setting, thereby permitting theoretical advancements in geometry and their scientific application to the world at large. "To those who held that space was finite [Archytas] issued the challenge that they explain to him why, were he taken to its edge, he could not reach beyond" (27). Archytas also held that place may preexist bodies and thereby constitute the first being (Jammer 1969, 11). Plato, in his *Timaeus,* identifies matter with the empty space of a "receptacle" or container in which forms become individual things. Now, if space is a receptacle, then it comes very close to being matter. However, Plato also sought to identify "the world of physical bodies with the world of geometric forms" (14), in other words, to link physics with geometry, or matter with representation and abstraction.

9. Reese (1980, 543) observes that Descartes's theorization of matter as infinite extension was a reversion to Parmenides' (515–450 B.C.) interpretation of *space* (also translated as *universe* in Kitto [1964, 182]) as a uniform, spherical plenum whose fullness made motion impossible. Parmenides rejected the Pythagorean void — a precursor of *absolute,* empty space — arguing that it was "nothing," and hence incapable of existence (Cornford 1936, 228).

10. In contrast to the space that is part of the substance of the world in which we live and that offers us existential grounding, *all* concepts of space (though admittedly coming into being in the minds of humans who are themselves situ-

ated in space) are abstract, representational, and partially the result of method—an issue pursued further in the discussion of space and metaphor in chapter 5. However, the phrase "abstract space" is not a tautology. There are degrees of abstraction that distinguish, for example, Newton's absolute space—which cannot be sensed—from his relative space and its association with sensation. In a geometric, abstract space, the nature or substance of objects is irrelevant. In a relative, abstract space, this is not the case, as human experience enters into play.

11. The principles of Euclidean geometry are invariant and unaffected by changes in substances. Euclidean geometry is ideal for conceptualizing both a self-subsisting representational space such as a VE and similarities between such a space and the natural world.

12. Virtually a guarantor of the concept of absolute space, Newton admitted a secondary role for what he identified as *relative* space, which is contained within absolute space. "Absolute space, in its own nature, without relation to anything external, remains always similar and immovable. Relative space is some movable dimension or measure of the absolute spaces; which our senses determine by its position to bodies; and which is commonly taken for immovable space; such is the dimension of a subterraneous, an aerial, or celestial space, determined by its position in respect of the earth. Absolute and relative space are the same in figure and magnitude; but they do not remain always numerically the same" (Newton 1946, 6).

Newton's relative space is arguably somewhat insignificant within the scope of his physics. It is nonetheless of interest here as he conceives it to be the *discernible measure* of absolute, immutable space. "Because the parts of space cannot be seen, or distinguished from one another by our senses, therefore in their stead we use sensible measures of them" (ibid., 8). By the device of relative space, Newton privileges an ideal and a way of knowing that humans can never sense. He writes: "I do not define time, space, place, and motion as being well known to all. Only I must observe that the vulgar conceive those quantities under no other notions but from the relations they bear to sensible objects. And thence arise certain prejudices, for the removing of which, it will be convenient to distinguish them into absolute and relative, true and apparent, mathematical and common" (6).

Koyré (1957, 160–61) suggests that Newton's distinction leads to absolute (or intelligible) space being opposed to "common-sense" or (sensible) space. Koyré further suggests that, for Newton, relative space is attached to, and moves with, the body or the thing in motion through absolute space. Relative space becomes an aspect of sensed events, processes, and experiences. Absolute space geometrically "subtends"—in the sense of extending or stretching under or beneath—relative space (162).

13. The ancient Greeks did not refer to the concept of place as implying an appellation or naming of a God. According to Jammer (1969, 28), the earliest connection between space/place and God is the use of the word "place" (*makom*) as a name for God in first-century-A.D. Palestinian Judaism. In Jewish cabalist thought—as expressed in the *gematria*—the name of God and the word "place" both are constituted by the number 186 (32).

14. Newton was aware of the conceptual difficulty that might arise from a void between objects in absolute space and denied that one body might act on another at a distance through a vacuum without the mediation of anything else. However, Newton also denied that gravitational pull could be an innate or essential component of matter. In the third of a series of four letters to his friend Richard Bentley, Newton wrote: "It is inconceivable that inanimate brute matter should, without mediation of something else which is not material, operate upon and affect other matter without mutual contact, as it must be if gravitation, in the sense of Epicurus, be essential and inherent in it.... That gravity should be innate, inherent, and essential to matter, so that one body may act upon another at a distance through a *vacuum*, without the mediation of anything else, by and through which their action and force may be conveyed from one to another, is to me so great an absurdity that I believe no man who has in philosophical matters a competent faculty in thinking can ever fall into it" (in Koyré 1957, 178–79, taken from Richard Bentley, *Works*, vol. 3 [London, 1838], letter 3, p. 211).

Koyré makes the point that because gravitation could not be attributed to matter, Newton's interpreters therefore ascribed it to divine power (ibid., 183).

15. Cyberspace is similar to Newtonian absolute space in that it is empty at the moment of its creation, and before it is occupied by data and other representations. Treating space as empty—as cyberspace and absolute space both do—is not, however, the same as theorizing that space precedes objects (a necessity "within" invented, representational cyberspace), or that two objects' relationship to each other, for example, creates space.

16. Jones notes the conflation of space with its representation as a measure. It is true that space is more easily conceptualized when understood as distance conceived in units of measure. Visual perception theorist James J. Gibson (1966) makes an implicit link between space and measure. He argues that airspaces between solids are permanent in shape though not solid in and of themselves. He suggests that this fact may have led theorists to extend the concept of rigidity and apply it to "so-called empty space, to assume that even the outer space between the stars is like the airspace of Euclidean geometry" (9). Although Gibson finds this extension dubious, when he then asserts the rigidity of "terrestrial space," he does so by the same process of conceptual extension from tool to re-

ality, stating that the "measuring stick of the earthbound surveyor is rigid, and the distances and angles of the terrain conform to the laws of Euclid" (9).

17. The confusion in language about concepts of space is apparent enough in commentators' explanations of Leibniz's understanding of space. Although I am here linking his understanding of it to a modern relational view of space, Leibniz defines space as *relative* because he believes it derives from *interrelations* of entities, and Jammer (1969, 118) argues that Leibniz's view of space as a system of *relations* is, in effect, a joining of kinematic relativity with the concept of absolute space.

18. "Cartesian space" is the term used by VE researchers to refer to the cyberspatial grid within which the virtual world takes place.

19. Anthropologist John Noyes (1992, 29) comments that mythical spaces depend on geometric strategies of reference and consistency. VEs have many of the hallmarks of mythical space.

20. One of the subsections of Jones's text is entitled "The Space Theater." It is an uncannily prescient description of the experiential, psychological, and phenomenological reality available within a VE. I contacted Jones to inquire when it was written. The passage dates from late 1974. Jones "was trying to evoke an experience of space that bridged our notions of inside and outside...and to make the reader aware of the essential role of consciousness in creating our metaphors of space and other aspects of 'physical reality,' for that matter" (Jones, personal correspondence via E-mail, 7 May 1995). The passage follows.

Imagine a visit to the theater in a carnival of space. In the auditorium you find yourself flying over the skyscrapers of New York. They erupt vertically toward you, separated by a rectangular grid of dizzying narrow valleys. It's like the opening of the film *West Side Story*, seen in 3-D wraparound. The buildings, like great rhomboidal crystals, sway ponderously and silently below you as your changing perspective has them tilt from the vertical, now this way, now that. The view of the rooftops, towers, and spires alternates with that of the streets far below, rapidly telescoping your perception of depth and height—up, down, in, out, high, low. Your gaze falls deeply into the space below. You sense the frightening penetrability of space, its unfathomable depths, its infinite reaches.

Suddenly, you are over the great canyons—Grand, Bryce. The extremes of height and depth still appear, but now, the rectangular regularity of the skyscrapers is replaced by organic, irregular shapes—buttes, mesas, pinnacles, cliffs. Views of the Alps and Himalayas follow, bringing with them a gradual heightening of depth perception and diminishing sense of proportion and size. Confusion increases as scenes alternate and overlap. Are these mountains you see or stalagmites? Are they microscopic crystalline growths or perhaps greatly enlarged organic tissue structures? Are you looking upward into the pendant vines and mosses of a tropical hanging garden, or down among the towering pines and redwoods of a coastal rain forest? Is that a morning mist permeating the air between the walls of a ravine or are you beneath the murky waters of the ocean, peering into an undersea gorge?...Do you really feel something? Is this a visual hallucination, a tactile illusion, or reality?...

You decide you've had enough. You get up and head for the exit. But the door seems to be a projection on the wall like the views above you. There doesn't seem to be an exit....You decide to sit on the floor and wait for the "show" to end. You're relieved to find that the floor is solid and not a projection....Must be some newfangled mental projector. What ever happened to the nice simple, clean space we all knew and loved so well, you wonder. It always knew its place and stayed there — outside of me. But the space in this theater is all somehow connected. There's no clear inside and outside. And it seems to be more than just space. It seems to be sensations and feelings as well.

...You discover, to your alarm...that there's no clear boundary between your inside and the outside! You fear that you're trapped in this strange tangled web, but you find that you can, in fact, move (if you can call it that). You no longer seem to have a body in the normal sense of the word. Rather, you are like a vortex or a pattern of concentration among all the flows and channels. To your amazement, it's all rather pleasant. You're not so much tangled up in things as you are connected to them. You are aware of thoughts, feelings, and experiences that are not your own, not necessarily even those of other people. When you wish to move, you pass effortlessly like a wave through the connective matrix. It's more like a thought moving through a mind than a body moving in space. Your former experience of space as an empty, geometrical void in which things have an existence as isolated entities, separated from each other by distance, has been transformed. You now experience a realm filled with meaning and wisdom, and in which things blend and participate in each other's being and significance. The projective, perspectival geometry of the artist and physicist has become the participatory connectivity of the alchemist and seer. Space is mind. Here meaning and information are not transported *across* space and *in* time. Instead, they are shared and omnipresent, for the fundamental relations are symbolic, rather than spatial. It is meaning which connects things, not distance which separates them. Causality is enlarged to synchronicity. This is a domain somewhere between the multiplicity of everyday consciousness and an ultimate unity (1982, 55–57, cited with permission of author).

21. When existence and space are reconceptualized as pure functions of communications, a need may then arise for an insecure human consciousness to communicate its existence and to render itself in material forms. This distinction between existence and consciousness is very close to one between nature and culture. Consciousness cannot exist without language and other symbolic representations that communications require — except unto itself. In this way, the virtual world and self-consciousness are similar. On its own, communication becomes a means to stave off uncertainty at the cost of only having reference to a world of symbols. The sphere of meaning, discourse, or language is made an access mechanism to the banished sphere of Being. However, need Kant's phenomenon — considered as a meeting point between the spheres of nature and society or culture, or between empirical reality and modern consciousness — therefore be thought of only as a "site" where nothing happens? Is the phenomenon not the encounter of already present elements (Latour 1993, 81)? Might it not also be understood as a spatio-temporality or "middle ground" where and when consciousness also experiences that it is in place? Further, might the phenomenon not be conceptually engaged to suggest a relationality between human agents and technology? Such an engagement would not reduce humanity to a deterministic clockwork mechanism, and neither would it situate technology as the "other" that we then come to loathe.

22. De Certeau's use of space here is in direct opposition to his theorizing of the meaning of "place" as official, monumental, and authoritarian.

23. I would like to expand the potential of "betweenness" to something more like "amongness" and thereby avert any inadvertent reduction of the agency of a place to no more than two "geographic individuals" present at any one time.

24. The importance of differences among places is distinct from the contemporary theorization of difference, which sometimes seems a defense of formal categories that ignores context or content. However inadvertently, a focus on categories is the same epistemological process that adjudicates places as empty containers awaiting received meaning. Containers generally are discrete. Any overlap among the "different containers" is not even considered.

25. It is the reactionary politics to which an aesthetics based on this kind of privileged use of bounded space can contribute that, in part, provokes Harvey's (1989, 206) objections to the academic use of aesthetic theories. However, landscape production is not reactionary per se, as Ann Jensen Adams's (1994) history of Dutch landscape painting's contribution to the rise of an anti-Hapsburg Dutch nationalism suggests.

26. Yet something like "the reverse" is also possible in a VE. Our bodies — part of the material world — become texts depicted within the landscape of VE. The sensual experience of the world is reduced to inscription, reading, and writing.

4. Sight and Space

1. Within digital computation, nothing can exist without a name. If I attempt to shut this document without naming it, the computer will prompt me for a name. If the power fails, a name will be assigned by the software. Otherwise, the file would have no *location* within the system. In other words, within cyberspace/computation/VR, name = space of existence. In magical terms, name becomes the "signature" for space. In magic, things take power from signature, incantation, and the casting of spells; in a VE, "things" cannot precede this action, and the resistant, prediscursive material reality of our bodies is not welcome, except, perhaps, as a "trace" from which we have achieved liberation.

In concrete reality, naming is also a paramount social power but remains secondary or temporally subsequent to the sensate existence of the flesh. Tuan (1990) notes that we tame by naming, that to name is part of placemaking. Existence precedes the quasi-magical powers of naming. We categorize the natural world through naming it, but the physical world preexists this action and has no existential requirement for it.

2. Biocca and Lanier (1992), but see also interview with W. Gibson, "Fantastic Voyages," *Macleans* 105, no. 50 (1992):44.

3. The San Francisco Audium — "A Theatre of Sound-Sculptured Space" — attempts to put in place a sound-space continuum. Stan Shaff, its creator-composer, writes that "sounds are 'sculpted' through their movement, direction, speed and intensity on multiple planes in space." He conceives of "sound as time and space; sound as object, environment or event" (Program Notes, Audium, San Francisco, Calif.).

4. Hence, ironically, it is ideal for colonization by VEs.

5. I concur with Jonas's observation that sight's ability to perceive things at a distance confers "tremendous biological advantage," that "knowledge at a distance is tantamount to foreknowledge." Where I might demur is when he then comments that sight's "*uncommitted* reach into space is gain of time for

adaptive behavior" (1982, 151; emphasis added). As Jonas has explained, "seeing requires no perceptible activity" by the subject or object. It seems effortless. It is also in the nature of sight to "withhold the experience of causality" (149). Jonas extends this argument to suggest that inasmuch as Hume looked no further than sight, he was correct to deny causal information. Had Hume linked vision with the other senses, he would have had to disavow his own theory. However, this argument by Jonas is still made within a framework that understands sight as "king" and the other senses as "subject" to his rule. It is this detachment of noble sight from the rest of its kingdom on the part of theorists as divergent as Jonas and James J. Gibson that allows belief that a sighted reach into space might somehow remain "uncommitted" and objective, even at the same time as Jonas feels compelled to assert that sight is the most easily fooled sense and requires the other senses serving as its vassals for any grip it might retain on "common sense."

6. This passage also fleshes out something of Jonas's theory as to why vision masks its causal connections.

7. Aristotle, *Physics,* 4.6.213b, 23. Cited in Cornford (1936, 223).

8. "The highly visual associations used by the Pythagoreans derive, some say, from the practice of setting forth sums by laying out pebbles on a smooth surface" (Reese 1980, 470). With reference to square numbers, Kitto (1964, 192) diagrams this association in the following way:

The statement "$1^2 + 3 = 2^2$" can be shown thusly

Similarly, the statement "$2^2 + 5 = 3^2$" can be represented as

And so forth . . .

9. Anaximander rejected earlier belief in the world as composed of a single primary element (most often water). Instead, the universe resulted from interactions of opposites. The Earth floated in air, columnar-like, within the Sun, Moon, and planets, which were theorized as rings of fire (Thuan 1995, 11).

10. Simplicius, *Physics,* 648, 11. Cited in Cornford (1936, 230).

11. I was struck in reading this passage with how Ellis's metaphor of the billiard cue ball for the self accords with Georg Simmel's account, in *The Philosophy of Money,* of the modern observer or subject as having come to be only constituted within flux and mobility.

12. Following Guy Debord (1994), one might assert that within a VE, the visionary self is made a spectacle in order to further enslave it within the panopticized virtual world and the commodification of reality this entails. However

(whatever the reasons behind an increasing stress level experienced by, and an increasing emphasis on the primacy of, the autonomous and fully individuated self), the fluidity being designed into these machines is premised in part on individual access. Previously more public self-performances are relocated to a "site" approximating the private sphere of interior light. Even if the goal of virtual technology is only the cynical manipulation of alienated selves anticipated by the first argument, the technology itself reflects the tenacity of widespread belief in this private self, along with the tension between relational and absolute conceptions of space.

13. The iconographic features of Macintosh computers or the PC Windows environment are rudimentary 2-D examples.

14. An interesting account of nineteenth-century concerns about the power of invisibility and anxieties over certainty is provided by Beer (1996).

15. Figure 6 also suggests something of NASA's efforts to model a schizophrenic space in which aspects of the subject coexist in spatial displays with other objects. NASA has sought to design a VE that would allow pilots conceptually to step out of their bodily referent and, depending on which point of view they have elected, either to turn back to gaze on their body/shell (which then becomes an object or other) or to sally forth in cyberspace toward a second representation of their body, which they would be able to penetrate iconically "at will" (Kroker 1992). This work on VEs is connected to NASA's interest in projecting telepresence and telerobotics. This is action at a distance, occurring in real time, and conforms to a thumbnail definition of magic. A robot in space, or in an environmentally contaminated site, would act as if sentient, because a human "presence," not actually inside the robot, could operate it at a distance by "interfacing" with its icon in a VE. Telepresence includes the scenario of deadly, apparently sentient robots, or killing machines, as part of the U.S. military's research agenda.

16. Denis Hollier (1984) relates how Jeremy Bentham, the panopticon's inventor, was tormented by a lifelong fear of ghosts. Fear of the invisible leads to the panopticon as a "built statement" that everything can be rendered visible. Indeed, Oettermann (1997, 353 n. 108) notes that Bentham had designed a residence for himself based on the device, suggesting that the transparency of the house of glass, or "glass architecture" of Scheerbart, is already at work.

17. This accords with my own experience of VR, noted in the introduction, which allowed me to fly imaginatively whenever I closed my eyes for several days after my experience within the VE had ended.

18. Hollier is citing from Caillois, "L'aridité," *Mesures*, April 1938.

19. No page number; citation taken from an article published in the electronic journal *Electronic Journal of Virtual Culture (EJVC)*.

20. I am grateful to Sean Smith for his comments on VR as a panoptical device. In late November 1996, a discussion "thread" on the listserv "Technology" — technology@lists.village.virginia.edu — to which both Smith and I belonged at the time, turned to potential relationships between the panopticon and contemporary IT and VR. I posted to the list a question addressing links between Foucault's understanding of the panopticon and the contemporary surveillance potential of VR. Smith's reply (20 November 1996) that "VR restricts the range of experiences from 'as many as there are people' to 'as many as there are different VR devices,' " and that VR helps construct consent because it "requires that we actively agree to take part in it, rather than the sometimes ambivalent viewing practices encouraged by radio and the telly" was useful in theorizing the interplay between pleasure and surveillance in VR.

21. Although the concept of progress today is widely denigrated, *technological* progress is widely accepted, even assumed and anticipated by otherwise critical academic theorists. Hence, social constructionists dismiss examinations of technology's agency as participating in a technological determinism that is the handmaiden of a mystifying diversion of attention away from the primacy of social relations. A naturalized belief in technological progress often assumes communications technologies as intermediary conduits through which pass unaltered messages between actors. Too ready a dismissal of nonhuman agencies partakes of the belief that since the apparent technical "conquest of nature," nothing beyond the human has affect in this world.

22. See, for example, Michael McCauley and Thomas Sharkey (1992, 312–13), who suggest that motion sickness can be considered an unwanted and perhaps inevitable side effect of traveling through VEs. They suggest that the "farther" the virtual travel, the greater the potential for nausea. "Both teleoperator and VE systems violate the normal correspondence between visual and vestibular patterns of sensation regarding self-motion."

23. George Berkeley, "Concerning the Principles of Human Knowledge," in *Toward a New Theory of Vision*, cited in Jammer 1969, 135.

24. The notion that extending one's hand out into space can be translated into an icon of this hand reaching out into the representational space of a VE derives from James J. Gibson's idea that we visually map or graft the dimensions of our world onto an internal perception-structuring system.

5. Space, Language, and Metaphor

1. The review of metaphors of light makes use of Blumenberg's 1957 history "Light as a Metaphor for Truth," published in English in 1993. For Blumenberg, the use of metaphor and narrative gives meaning to what would otherwise be a meaningless existence. Indeed, the philosophy and history of thought cannot

be detached from metaphoric language; at base, he seems to say, concepts are metaphors whose origins we have forgotten.

2. The ancient notion of light as a "wherein" that precedes the materiality to which it gives illumination and therefore spatial relations is not so very different from modern acknowledgments of the *basic* nature of light, and the modern theory of light, which can be stated only in mathematical form (Brill 1980, 4). Brill notes that because of this, "rather than trying to go further in 'explaining' light, it is more useful to concentrate on its practical properties" (4) — a statement in which epistemology swallows ontology and not unlike assertions by some geographers, who, in their interest in studying how "space" is used, forget the politics that always freight its various conceptualizations and how they are then put into discourse.

3. W. Brocker, *Aristoteles* (Frankfurt, 1935), 148. Cited in Blumenberg (1993, 55 n. 15).

4. Kitto (1964, 194) makes a similar observation to Taylor's: "Although Plato does not formally identify the Good with God, he speaks of its divine nature in such a way that formal identification would make but little difference."

5. Blumenberg (1993, 46) describes *logoi*, for the Greeks, as a collection of what had been seen. Logic was a general theory of the ways a "mental product," idea, or concept reflected or mirrored the world truthfully. Kitto (1964, 187) notes that *logos* is usually mistranslated as "word" but means speech or the "idea conveyed by speech." From this, one might extrapolate to suggest that the phrase "In the beginning was the Word" could be taken to mean "In the beginning was the visual concept."

6. Augustine, *De ordine*, 15.42. Cited in Hofstadter and Kuhns 1976, 180.

7. This self-protective and insular medieval cultural move is not unlike that taken by many contemporary subjects choosing virtuality over reality.

8. Koyré (1957, 16) translates Nicholas of Cusa's comments about truth and the relativity of spatial perception. As no one place "can claim an absolutely privileged value ... we have to admit the possible existence of different, equivalent world-images, the relative — in the full sense of the world — character of each of them, and the utter impossibility of forming an objectively valid representation of the universe." Translated from *De docta ignorantia* 2.2.99.

9. The word "*luxury*" contains a justification for the status quo of social inequalities and metropolitan privilege. Luxury connotes the "natural" stomping ground that is the due of those whose *lux* best reflects divine il*lumen*ation.

10. The quotation is from episode 1, "Blipverts," 1985.

11. Enter Jaron Lanier's wish for a virtual world of "post-symbolic communication" (Biocca and Lanier 1992, 161). Lanier's wish is an expression of the insecurity attending how we operate as moral agents in the world. The cultural

push toward all things virtual is a magical yearning that echoes cabalist assertions that a universal harmony is achievable through sound, shape, and number.

12. Both scenarios suggest a merger of subjectivity and space. Because cyberspace and VEs are based on language and code, Lefebvre's critique of structuralist semiotics' theorization of space as given with and in language is worth noting. "[Space] is not formed separately from language. Filled with signs and meanings, *an indistinct intersection point of discourses, a container homologous with whatever it contains,* space so conceived is comprised merely of functions, articulations and connections — in which respect it closely resembles discourse" (1992, 136; emphasis added).

13. Virtual technologies have the related potential to supersede ideology in their ability to reconstitute the subjectively held myths of social institutions as digital information.

14. The meanings of humans bodies are reduced to digital information. Further, the idea that corporeal presence is no more than an information bundle that can be expressed satisfactorily through iconographics fosters a belief that everything we need to know about our bodies can be expressed by a cartoon.

15. Benjamin Whorf comments that all European languages try "to make time and feelings visible, to constrain them to possess spatial dimensions that can be pointed to, if not measured." From Whorf, *Collected Papers on Metalinguistics* (Washington, D.C.: Foreign Service Institute, Department of State, 1952). Cited in Tuan 1977, 393. Whorf's comment is controversial in that the broader substance of his theory linking the structure of language to how humans think has been challenged by the more cybernetic, "hardwired" approach taken to language by Steven Pinker (1994, 60–61).

6. Identity, Embodiment, and Place

1. Haraway embraces the cyborg for its political possibilities precisely because she adjudicates its developmental and cultural trajectories as inevitable. She would rather be a cyborg than a goddess, for it is her project to introject herself into the masculinist scientific enterprise and thereby work toward reforming it from within — a position she perceives to be more effective than alternative strategies that avoid the implications of science and technology until it is too late. The exterior opponents are then enmeshed, for example, by the cyborg, and by other dominant strategies issuing from science and technology, without having had a say in their development.

2. Transcription from personal notes. David Levin, "Existentialism at the End of Modernity," lecture, Programme of Social and Political Thought, York University, Toronto, 3 April 1991.

3. Barker (1984, 63) notes that when body and subject (or text) separate, so too do desire and meaning. After this we come to have the opinion as subjects that desire is meaningless and that meaning does not relate to desire. Virtual bodies in virtual worlds are an attempt at simulating an ersatz rejoining of this sundered unity.

4. Downloaded from the Geo-Ethics listserv, Geo-Ethics@atlas.socsci. umn.edu. Posted to the list Thursday, 12 June 1997.

5. Disembodiment (Gardner's "Invisible Cameramen" noted in chapter 4) may be felt to occur when one's body is marginalized during virtual subjective experience. However, it is prudent to acknowledge that any such claim may be potentially nostalgic to the degree that it is not really a noting of actual body loss but perhaps only of a subjective ideation of the human body that always did lie at a certain conceptual remove from the modern subject's location.

6. Dazzlement has the power to elicit active agreement on the part of the subject. I find the dynamic of VEs coercive, but I acknowledge the saliency of Antonio Gramsci's (1988, 401) statement that "coercion is such only for those who reject it."

7. This is the position of figure A2 in figure 6, or of the subject looking through a window in Elias's modern spatial wall.

8. Augustinian Neoplatonism and the Enlightenment accord the individual an agency, which includes the power to position oneself as looking into the light. With VEs, however, this agency is tempered by the computer's power to act as a godlike panoptic eye, and to suggest that no matter where we look or turn, its sentient power has been there before us. This assumes the issue of lag can be resolved. This delay of one-thirtieth of a second between user movement and computational ability to update the visual display not only creates disorientation for some users but also may permit users to position themselves more critically vis-à-vis the machine, calling to attention, as it does, the constructedness of the scene along with the "periphery" of a user's visual attention. Virtual technology research is confident this lag will be overcome soon. Whether one greets this with anticipation or dismay is a separate issue.

9. As the critical history of VR in chapter 1 suggests, it is not inconceivable for scientific knowledge to be combined with transcendent desire to authorize production of a seemingly magical space. To think of this space as benign is an error, however, one that the U.S. military has not made — as its increasing demand for sophisticated simulations of killing fields and battle zones attests.

10. Ironically, perhaps, this is already somewhat implied in figure 7's depiction of the militarized male subject who customizes his virtual world in advance of entry.

11. "Means and *Conduits*" are intended to dilute the conscience and religious fervor of the competitive individual. Conduits dilute this fervor into the form of "opinions" ultimately represented by the king. In this, I would suggest, Hobbes anticipates an understanding that communication technology is *not* value neutral.

12. "Telepresence offers the ability to remotely manipulate physical reality in real time through its image. The body of a teleoperator is transmitted, in real time, to another location where it can act on the subject's behalf: repairing a space station, doing underwater excavation or driving a toy vehicle over the Nibelungen Bridge" (Manovich 1995b).

13. Victor Tausk, "On the Origin of the 'Influencing Machine' in Schizophrenia," *Psychoanalytic Quarterly* 2 (1933 [1919]), cited and reviewed in Sass 1994.

14. Any writer using word processing will know the horror of losing a section of work he or she has not saved, which may thereby be lost during a crash or malfunction. At such moments, one can sense how aspects of one's intelligence or psyche have been exteriorized, in this case transferred, into the machine.

15. The suggestion that VEs might eliminate the need for carpentry skills partakes of the "blind belief" Mazlish (1967) notes in "machines' abilities to solve all problems." Such a belief dovetails with a rampant optimism that welcomes technological benefits but assumes they come with few or no social and material costs. The fiction of cost-free technical innovation (I am reminded of the nuclear power industry's slogan of a generation ago — "too cheap to meter") is, at base, a radically conservative one in that it suggests things will not change. Somehow technology is to magically bestow its benefits while social and material relations remain unaltered.

16. Taken from Casey 1987, 153. Emphasis added. Cited in Igor Stravinsky, *The Poetics of Music* (New York: Random House, 1960), 68.

17. An earlier iteration of VR's extension of the landscape idea and how it relates to political formations is at work in the panorama. "After access to nature has been bought up by the propertied classes, the propertyless are permitted 'visual appropriation' in return for a small fee. . . . the panorama presents, blurs, and idealizes the circumstances of land ownership" (Oettermann 1997, 47).

18. As Morse (1994, 163) notes, much discussion centered on "the future" and VEs is hostile to organic life. Expressive intelligence that extends beyond the body's physical limits is seen not as organic but rather as joining with "data." The hopes for transcendence reflected in an emphasis on immersion seem to be a desire for *inorganic* rebirth, and they trade in a psychotic and fatal form of reasoning.

19. Equating space with distance leaves the subject in the oddly untenable position of conceiving his or her own existence as a standing out or emerging from distance. When the equation space = distance is replaced by space = move-

ment, it no longer seems to matter as much that the subject emerges in a dialectical relationship with space as long as she or he is always "on the go" — to the point that her or his form begins to lose integrity. I can only here suggest the value of further theoretical investigation into the emerging relationships among space as movement, this loss of integrity, and the fact that dreams of immortality are concerned less with what form we will take in the future than with asserting in advance the "fact" that mind will continue to persist.

References

Agnew, John. 1989. "The Devaluation of Place in Social Science." In *The Power of Place: Bringing Together Geographical and Sociological Imaginations*, ed. John Agnew and James Duncan. Boston: Unwin Hyman.

Aitken, Stuart, and Andrea Westersund. 1996. "Just Passing Through: Virtual Tourism, Justice, and 'Informatics.'" In *GIS and Society: The Social Implications of How People, Space, and Environment Are Represented in GIS, Scientific Report for the Initiative 19 Specialist Meeting*. Technical Report 96-7. Santa Barbara, Calif., and Buffalo, N.Y.: National Center for Geographic Information and Analysis.

Anchor, Robert. 1985. "Bakhtin's Truths of Laughter." *Clio* 14, no. 3.

Anderson, Benedict. 1991. *Imagined Communities: Reflections on the Origin and Spread of Nationalism*. Rev. ed. London: Verso.

Ardener, Shirley, ed. 1981. *Women and Space: Ground Rules and Social Maps*. London: Croom Helm.

Arendt, Hannah. 1958. *The Human Condition*. Chicago: University of Chicago Press.

Bakhtin, Mikhail. 1984. *Rabelais and His World*. Trans. Hélène Iswolsky. Bloomington: Indiana University Press.

Bal, Mieke. 1994. *On Meaning-Making: Essays in Semiotics*. Sonoma, Calif.: Polebridge Press.

Barfield, Owen. 1977. *The Rediscovery of Meaning and Other Essays*. Middleton, Conn.: Wesleyan University Press.

Barker, Frances. 1984. *The Tremulous Private Body*. London: Methuen.

Beaudry, Jean-Louis. 1974–1975. "Ideological Effects of the Basic Cinematographic Apparatus." Trans. Alan Williams. *Film Quarterly* 28, no. 2 (winter).

Beck, Ulrich. 1992. *Risk Society: Towards a New Modernity*. Trans. Mark Ritter. London: Sage.

Beer, Gillian. 1996. "'Authentic Tidings of Invisible Things': Vision and the Invisible in the Later Nineteenth Century." In *Vision in Context: Historical and Contemporary Perspectives on Sight*, ed. Theresa Brennan and Martin Jay. New York: Routledge.

Benedikt, Michael. 1992a. "Cyberspace: Some Proposals." In *Cyberspace, First Steps*, ed. Michael Benedikt. Cambridge, Mass.: MIT Press.

———. 1992b. Introduction to *Cyberspace, First Steps*, ed. Michael Benedikt. Cambridge, Mass.: MIT Press.

Bengtson, Hermann. 1988. *History of Greece: From the Beginnings to the Byzantine Era*. Trans. Edmund F. Bloedow. Ottawa: University of Ottawa Press.

Benjamin, Walter. 1979. *One-Way Street and Other Writings*. Trans. Edmund Jephcott and Kingsley Shorter. London: NLB.

Biocca, Frank. 1988. "The Pursuit of Sound: Radio, Perception, and Utopia in the Early Twentieth Century." *Media, Culture, and Society* 10.

———. 1992a. "Communication within Virtual Reality: Creating a Space for Research." *Journal of Communications* 42, no. 4.

———. 1992b. "Virtual Reality Technology: A Tutorial." *Journal of Communications* 42, no. 4.

———. 1992c. "Will Simulation Sickness Slow Down the Diffusion of Virtual Environment Technology?" *Presence* 1, no. 3.

Biocca, Frank, and Jaron Lanier. 1992. "An Insider's View of the Future of Virtual Reality." *Journal of Communications* 42, no. 4.

Blondheim, Menachim. 1994. "When Bad Things Happen to Good Technologies: Three Phases in the Diffusion and Perception of American Telegraphy." In *Technology, Pessimism, and Postmodernism*, ed. Y. Ezrahi et al. Dordrecht, Netherlands: Kluwer Academic Publishers.

Blumenberg, Hans. 1993. "Light as a Metaphor for Truth: At the Preliminary Stage of Philosophical Concept Formation." Trans. Joel Anderson. In *Modernity and the Hegemony of Vision*, ed. David Michael Levin. Berkeley: University of California Press.

Boden, Deirdre, and Harvey L. Molotch. 1994. "The Compulsion of Proximity." In *NowHere: Space, Time, and Modernity*, ed. Roger Friedland and Deirdre Boden. Berkeley: University of California Press.

Bolter, Jay David. 1984. *Turing's Man: Western Culture in the Computer Age*. Chapel Hill: University of North Carolina Press.

———. 1991. *Writing Space: The Computer, Hypertext, and the History of Writing*. Hillsdale, N.J.: Lawrence Erlbaum.

Bordo, Susan. 1993. "'Material Girl': The Effacements of Postmodern Culture." In *The Madonna Connection: Representational Politics, Subcultural*

Identities, and Cultural Theory, ed. Cathy Schwichtenberg. Boulder, Colo.: Westview Press.

Bourdieu, Pierre. 1984. *Distinction: A Social Critique on the Judgment of Taste.* Trans. Richard Nice. Cambridge, Mass.: Harvard University Press.

Brand, Stewart. 1987. *The Media Lab: Inventing the Future at MIT.* New York: Viking Penguin.

Brenneman, Walter L., Jr., Stanley Yarian, and Alan Olson. 1982. *The Seeing Eye: Hermeneutical Phenomenology in the Study of Religion.* University Park: Pennsylvania State University Press.

Brewster, David. 1832. *"Letters on Natural Magic," Addressed to Sir Walter Scott, Bart.* London: John Murray.

———. 1856. *The Stereoscope: Its History, Theory, and Construction.* London: John Murray.

Brill, Thomas B. 1980. *Light: Its Interaction with Art and Antiques.* New York: Plenum Press.

Brooks, Fred. 1991. "Statement of Dr. Fred Brooks." In *New Developments in Computer Technology: Virtual Reality.* Hearing before the Subcommittee on Science, Technology, and Space of the Committee on Commerce, Science, and Transportation, United States Senate, S. Hrg. 102-553, 8 May.

Brown, Norman O. 1966. *Love's Body.* New York: Random House.

Bryson, Norman. 1983. *Vision and Painting: The Logic of the Gaze.* New Haven, Conn.: Yale University Press.

Buck-Morss, Susan. 1989. *The Dialectics of Seeing: Walter Benjamin and the Arcades Project.* Cambridge, Mass.: MIT Press.

———. 1993. "Dream World of Mass Culture." In *Modernity and the Hegemony of Vision,* ed. David Michael Levin. Berkeley: University of California Press.

Bukatman, Scott. 1993. *Terminal Identity: The Virtual Subject in Post-Modern Science Fiction.* Durham, N.C.: Duke University Press.

Bunn, David. 1994. " 'Our Wattled Cot': Mercantile and Domestic Space in Thomas Pringle's African Landscapes." In *Landscape and Power,* ed. W. J. T. Mitchell. Chicago: University of Chicago Press.

Burgin, Victor. 1989. "Geometry and Abjection." In *The Cultural Politics of Postmodernism,* ed. John Tagg. Binghamton, N.Y.: Department of Art and Art History, SUNY Binghamton.

Bush, Vannevar. 1946. *Endless Horizons.* Washington, D.C.: Public Affairs Press.

Caillois, Roger. 1984. "Mimicry and Legendary Psychasthenia." Trans. John Shepley. *October* 31 (winter).

Campbell, Jeremy. 1982. *Grammatical Man: Information, Entropy, Language, and Life.* New York: Simon and Schuster.

Carey, James W. 1975. "A Cultural Approach to Communication." *Communication 2*.

———. 1983. "Technology and Ideology: The Case of the Telegraph." *Prospects 9*.

———. 1989. *Communication as Culture*. Boston: Unwin Hyman.

Carey, James W., and John J. Quirk. 1969–1970. "The Mythos of the Electronic Revolution." *American Scholar* (winter).

Carroll, Noel. 1988. *Mystifying Movies: Fads and Fallacies in Contemporary Film Theory*. New York: Columbia University Press.

Carruthers, Mary. 1990. *The Book of Memory: A Study of Memory in Medieval Culture*. Cambridge: Cambridge University Press.

Casey, Edward S. 1987. *Remembering: A Phenomenological Study*. Bloomington: Indiana University Press.

Castle, Terry. 1995. *The Female Thermometer: Eighteenth-Century Culture and the Invention of the Uncanny*. New York: Oxford University Press.

Cavalcanti, Petro, and Paul Piccone. 1975. *History, Philosophy, and Culture in the Young Gramsci*. St. Louis, Mo.: Telos Press.

Cavell, Stanley. 1971. *The World Viewed*. New York: Viking.

Chang, Briankle G. 1996. *Deconstructing Communication: Representation, Subject, and Economies of Exchange*. Minneapolis: University of Minnesota Press.

Connor, James A. 1993. "Strategies for Hyperreal Travelers." *Science Fiction Studies* 20, part 2 (March).

Copleston, Frederick. [1960] 1994. *A History of Philosophy*. Vol. 6. New York: Image Books.

Cornford, Francis Macdonald. 1936. "The Invention of Space." In *Essays in Honour of Gilbert Murray*, ed. James Alexander Kerr Thomson and Arnold Joseph Toynbee. London: George Allen and Unwin.

Cosgrove, Denis. 1984. *Social Formation and Symbolic Landscape*. London: Croom Helm.

Couch, Herbert Newell, and Russell M. Geer. [1940] 1961. *Classical Civilization, Greece*. 2d ed. Englewood Cliffs, N.J.: Prentice-Hall.

Coyne, Richard. 1994. "Heidegger and Virtual Reality." *Leonardo* 27, no. 1.

Crary, Jonathan. 1994. *Techniques of the Observer: On Vision and Modernity in the Nineteenth Century*. Cambridge, Mass.: MIT Press.

Cronon, William. 1991. *Nature's Metropolis*. New York: W. W. Norton.

Curry, Michael. 1995. "GIS and the Inevitability of Ethical Inconsistency." In *Ground Truth: The Social Implications of Geographic Information Systems*, ed. John Pickles. Guilford, N.Y.

———. 1996. "On Space and Spatial Practice in Contemporary Geography." In *Concepts in Human Geography*, ed. Carville Earle et al. Lanham, Md.: Rowman and Littlefield.

Daniels, Stephen. 1993. *Fields of Vision: Landscape Imagery and National Identity in England and the United States.* Cambridge: Polity Press.

Davidson, Donald. 1978. "What Metaphors Mean." In *On Metaphor,* ed. Sheldon Sacks. Chicago: University of Chicago Press.

Davis, Erik. 1993. "Techgnosis, Magic, Memory." In *Flame Wars: The Discourse of Cyberculture,* ed. Mark Dery. Durham, N.C.: Duke University Press.

Debord, Guy. 1994. *The Society of the Spectacle.* Trans. Donald Nicholson-Smith. New York: Zone.

de Cauter, Lieven. 1993. "The Panoramic Ecstasy: On World Exhibitions and the Disintegration of Experience." *Theory, Culture, and Society* 10.

de Certeau, Michel. 1984. *The Practice of Everyday Life.* Trans. Steven Rendall. Berkeley: University of California Press.

de Kerckhove, Derrick. 1991. "The New Psychotechnologies." In *Communication in History: Technology, Culture,* Society, ed. David Crowley and Paul Heyer. White Plains, N.Y.: Longman.

Deleuze, Gilles, and Félix Guattari. 1987. *A Thousand Plateaus: Capitalism and Schizophrenia.* Trans. Brian Massumi. Minneapolis: University of Minnesota Press.

Department of Computer Science, University of North Carolina at Chapel Hill. 1997. *Virtual Backdrops: Replacing Geometry with Textures.* Department of Computer Science, University of North Carolina at Chapel Hill.

Depew, David J. 1985. "Narrativism, Cosmopolitanism, and Historical Epistemology." *Clio* 14, no. 4.

Dery, Mark. 1993a. *Culture Jamming: Hacking, Slashing, and Sniping in the Empire of Signs.* Westfield, N.J.: Open Media.

———. 1993b. "Flame Wars." In *Flame Wars: The Discourse of Cyberculture,* ed. Mark Dery. Durham, N.C.: Duke University Press.

Dewey, John. 1916. *Democracy and Education.* New York: Macmillan.

Division Incorporated. 1994. "Weapon Systems in Virtual Trials." Application brief. Redwood City, Calif.

DiZio, Paul, and James R. Lackner. 1992. "Spatial Orientation, Adaptation, and Motion Sickness in Real and Virtual Environments." *Presence* 1, no. 3.

Dobson, Jerome E. 1993. "The Geographic Revolution: A Retrospective on the Age of Automated Geography." *Professional Geographer* 45, no. 4.

Dretske, Fred I. 1969. *Seeing and Knowing.* London: Routledge and Kegan Paul.

Dreyfus, Hubert L. 1992. *What Computers Still Can't Do: A Critique of Artificial Reason.* Cambridge, Mass.: MIT Press.

Dreyfus, Hubert L., and Patricia Allen Dreyfus. 1964. Translator's introduction to *Sense and Non-Sense,* by Maurice Merleau-Ponty. Evanston, Ill.: Northwestern University Press.

Eco, Umberto. 1983. "Travels in Hyperreality." In *Travels in Hyperreality,* trans. William Weaver. San Diego, Calif.: Harcourt, Brace, Jovanovich.

Edgerton, Samuel Y., Jr. 1975. *The Renaissance Rediscovery of Linear Perspective.* New York: Basic Books.

Edwards, Paul. 1989. "The Closed System: Systems Discourse, Military Policy, and Post–World War II US Historical Consciousness." In *Cyborg Worlds: The Military Information Society,* ed. Les Levidow and Kevin Robins. London: Free Association Books.

Einstein, Albert. 1969. Foreword to *Concepts of Space: The History of Theories of Space in Physics,* by Max Jammer. 2d ed. Cambridge, Mass.: Harvard University Press.

Elias, Norbert. 1968. *The Civilizing Process: The History of Manners.* Trans. Edmund Jephcott. New York: Urizen.

Ellis, Stephen. 1991a. "Pictorial Communication: Pictures and the Synthetic Universe." In *Pictorial Communication in Virtual and Real Environments,* ed. Stephen Ellis. New York: Taylor and Francis.

———. 1991b. Prologue to *Pictorial Communication in Virtual and Real Environments,* ed. Stephen Ellis. New York: Taylor and Francis.

Emery, Fred, and Merrelyn Emery. 1976. *A Choice of Futures.* Leiden, Netherlands: Martinus Nijhoff.

Entrikin, Nicholas. 1991. *The Betweenness of Place: Towards a Geography of Modernity.* Baltimore, Md.: Johns Hopkins University Press.

Feenberg, Andrew. 1991. *Critical Theory of Technology.* New York: Oxford University Press.

Ferré, Frederick. 1995. *Philosophy of Technology.* Athens: University of Georgia Press.

Fiske, John. 1988. "Popular Forces and the Culture of Everyday Life." *Southern Review* 21.

———. 1991. "Postmodernism and Television." In *Mass Media and Society,* ed. James Curran and Michael Gurevitch. London: Edward Arnold.

———. 1992. "The Culture of Everyday Life." In *Cultural Studies,* ed. Lawrence Grossberg, Cary Nelson, and Paula A. Treichler. New York: Routledge.

Fitting, Peter. 1991. "The Lessons of Cyberpunk." In *Technoculture,* ed. Constance Penley and Andrew Ross. Minneapolis: University of Minnesota Press.

Foucault, Michel. 1970. *The Order of Things.* New York: Vintage Books.

———. 1979. *Discipline and Punish: The Birth of the Prison.* New York: Vintage Books.

———. 1988. *Technologies of the Self: A Seminar with Michel Foucault.* Ed. Luther Martin, Huck Gutman, and Patrick Hutton. Amherst: University of Massachusetts Press.

Friedberg, Anne. 1993. *Window Shopping: Cinema and the Postmodern.* Berkeley: University of California Press.

Gane, Mike. 1988. *On Durkheim's Rules of Sociological Method.* London: Routledge.

Gardner, Brian. 1993. "The Creator's Toolbox." In *Virtual Reality: Applications and Explorations,* ed. Alan Wexelblat. Boston: Academic Press Professional.

Gelernter, David. 1992. *Mirror Worlds: Or the Day Software Puts the University in a Shoebox.* New York: Oxford University Press.

Gibson, James J. 1950. *The Perception of the Visual World.* Boston: Houghton Mifflin.

———. 1966. *The Senses Considered as Perceptual Systems.* Boston: Houghton Mifflin.

Gibson, William. 1984. *Neuromancer.* New York: Ace Publishing.

———. 1986. *Count Zero.* New York: Ace Publishing.

———. 1988. *Mona Lisa Overdrive.* New York: Bantam Books.

———. 1992. "Academy Leader." In *Cyberspace, First Steps,* ed. Michael Benedikt. Cambridge, Mass.: MIT Press.

Gigliotti, Carole A. 1993. *Aesthetics of a Virtual World: Ethical Issues in Interactive Technological Design.* Ph.D. diss., Ohio State University.

Godwin, Joscelyn. 1979. *Athanasius Kircher: A Renaissance Man and the Quest for Lost Knowledge.* London: Thames and Hudson.

Goldhill, Simon. 1996. "Refracting Classical Vision: Changing Cultures of Viewing." In *Vision in Context: Historical and Contemporary Perspectives on Sight,* ed. Theresa Brennan and Martin Jay. New York: Routledge.

Gramsci, Antonio. 1988. *An Antonio Gramsci Reader.* Ed. David Forgacs. New York: Schocken Books.

Grant, Glenn. 1990. "Transcendence through Detournement in William Gibson's *Neuromancer.*" In *Science Fiction Studies* 17, no. 50.

Grosz, Elizabeth. 1994. *Volatile Bodies.* Bloomington: Indiana University Press.

———. 1995. *Space, Time, and Perversion.* New York: Routledge.

Haraway, Donna. 1985. "A Manifesto for Cyborgs: Science, Technology, and Socialist Feminism in the 1980s." *Socialist Review* 15, no. 2.

———. 1991. "The Promises of Monsters." In *Cultural Studies,* ed. Lawrence Grossberg, Cary Nelson, and Paula Treichler. New York: Routledge.

Harrison, Charles. 1994. "The Effects of Landscape." In *Landscape and Power,* ed. W. J. T. Mitchell. Chicago: University of Chicago Press.

Hartley, John. 1992. *The Politics of Pictures: The Creation of the Public in the Age of Popular Media.* London: Routledge.

Hartshorne, Richard. 1959. *Perspectives on the Nature of Geography.* Chicago: Rand McNally.

Harvey, David. 1989. *The Condition of Postmodernity: An Enquiry into the Origins of Cultural Change*. Oxford: Basil Blackwell.

Havelock, Eric. 1982. *The Literate Revolution in Greece and Its Cultural Consequences*. Princeton, N.J.: Princeton University Press.

Hayles, N. Katherine. 1992. "The Materiality of Informatics." *Configurations* 1.

————. 1993a. "The Seductions of Cyberspace." In *Rethinking Technologies*, ed. Verena Andermatt Conley. Minneapolis: University of Minnesota Press.

————. 1993b. "Virtual Bodies and Flickering Signifiers." *October* 66 (fall).

————. 1995. "Simulated Nature and Natural Simulations: Rethinking the Relation between the Beholder and the World." In *Uncommon Ground: Toward Reinventing Nature*, ed. William Cronon. New York: W. W. Norton.

Heidegger, Martin. 1977. *The Question Concerning Technology and Other Essays*. Trans. William Lovitt. New York: Garland.

Heilig, Morton. 1992a. "El Ciné del Futuro: The Cinema of the Future." Trans. Uri Feldman. *Presence* 1, no. 3.

————. 1992b. "Enter the Experiential Revolution: A VR Pioneer Looks Back to the Future." In *Cyberarts: Exploring Arts and Technology*, ed. Linda Jacobson. San Francisco: Miller Freeman.

Heim, Michael. 1993. *The Metaphysics of Virtual Reality*. New York: Oxford.

Held, Richard. 1972. "Plasticity in Sensory-Motor Systems." In *Perception: Mechanisms and Models*, ed. Richard Held and Whitman Richards. San Francisco: W. H. Freeman.

Held, Richard, and Nathaniel Durlach. 1991. "Telepresence, Time Delay, and Adaptation." In *Pictorial Communication in Virtual and Real Environments*, ed. Stephen Ellis. New York: Taylor and Francis.

Hillis, Ken. 1994a. "The Power of Disembodied Imagination: Perspective's Role in Cartography." *Cartographica* 31, no. 3.

————. 1994b. "The Virtue of Becoming a No-Body." *Ecumene* 1, no. 2.

————. 1998. "Human.language.machine." In *Mapping the Body*, ed. Steve Pile and Heidi Nast. London: Routledge.

Hirschkop, Ken. 1991. "Is Dialogism for Real?" *Social Text* 30.

Hobbes, Thomas. [1651] 1985. *Leviathan*. Ed. Crawford Brough MacPherson. London: Penguin Books.

Hobsbawm, Eric. 1990. *Nations and Nationalism since 1780: Programme, Myth, Reality*. Cambridge: Cambridge University Press.

Hofstadter, Albert, and R. Kuhns, eds. 1976. *Philosophies of Art and Beauty*. Chicago: University of Chicago Press.

Hollier, Denis. 1984. "Mimesis and Castration 1937." Trans. William Rodarmor. *October* 31 (winter).

Idhe, Don. 1991. *Instrumental Realism: The Interface between Philosophy of Science and Philosophy of Technology.* Bloomington: Indiana University Press.

Innis, Harold. 1951. *The Bias of Communications.* Toronto: University of Toronto Press.

Jackson, Peter. 1996. "The Idea of Culture: A Response to Don Mitchell." *Transactions of the Institute of British Geographers* 21, no. 3.

Jameson, Frederic. 1984. "Postmodernism, or the Cultural Logic of Late Capitalism." *New Left Review,* no. 146.

Jammer, Max. 1969. *Concepts of Space: The History of Theories of Space in Physics.* 2d ed. Cambridge, Mass.: Harvard University Press.

Jensen Adams, Ann. 1994. "Competing Communities in the 'Great Bog of Europe': Identity and Seventeenth-Century Dutch Landscape Painting." In *Landscape and Power,* ed. W. J. T. Mitchell. Chicago: University of Chicago Press.

Johnston, Ronald, Derek Gregory, and David Smith. 1994. *The Dictionary of Human Geography.* 3d ed. London: Blackwell.

Jonas, Hans. 1982. *The Phenomenon of Life.* Chicago: University of Chicago Press.

Jones, Roger S. 1982. *Physics as Metaphor.* Minneapolis: University of Minnesota Press.

Jones, Steve. 1993. "A Sense of Space: Virtual Reality, Authenticity, and the Aural." *Critical Studies in Mass Communication* 10.

Kaiser, Mary. 1991. "Knowing." In *Pictorial Communication in Virtual and Real Environments,* ed. Stephen Ellis. New York: Taylor and Francis.

Katz, Ephraim. 1994. *The Film Encyclopedia.* New York: HarperPerennial.

Kelly, Kevin. 1994. "Hive Mind." *Whole Earth Review* 82.

Kendrick, Michelle. 1996. "Cyberspace and the Technological Real." In *Virtual Realities and Their Discontents,* ed. Robert Markley. Baltimore, Md.: Johns Hopkins University Press.

Kitto, Humphrey Davy Findley. 1964. *The Greeks.* Harmondsworth: Penguin.

Kolakowski, Leszek. 1990. *Modernity on Endless Trial.* Chicago: University of Chicago Press.

Koyré, Alexandre. 1957. *From the Closed World to the Infinite Universe.* Baltimore, Md.: Johns Hopkins University Press.

Krauss, Rosalind E. 1994. *The Optical Unconscious.* Cambridge, Mass.: MIT Press.

Kroker, Arthur. 1992. "Virtual Reality." Radio interview, *Turning Point,* La Bande Magnetique, Montreal, summer.

Kroker, Arthur, and Michael A. Weinstein. 1994. *Data Trash: The Theory of the Virtual Class.* New York: St. Martin's Press.

Krueger, Myron W. 1991a. "Artificial Reality: Past and Future." In *Virtual Reality: Theory, Practice, and Promise,* ed. Sandra K. Helsel and Judith Paris Roth. Westport, Conn.: Meckler.

———. 1991b. *Artificial Reality II.* Reading, Mass.: Addison-Wesley.

Kurland, Michael. 1984. "Of Gods, Humans, and Machines." In *Digital Deli,* ed. Steve Ditlea. New York: Workman.

Langer, Suzanne. 1985. "Discursive and Presentational Forms." In *Semiotics: An Introductory Anthology,* ed. Robert E. Innis. Bloomington: Indiana University Press.

Lash, Scott, and Brian Wynne. 1992. Introduction to *Risk Society: Towards a New Modernity,* by Ulrich Beck, trans. Mark Ritter. London: Sage.

Latour, Bruno. 1993. *We Have Never Been Modern.* Trans. Catherine Porter. Cambridge, Mass.: Harvard University Press.

Laurel, Brenda. 1991. *Computers as Theatre.* Reading, Mass.: Addison-Wesley.

Lefebvre, Henri. 1992. *The Production of Space.* Trans. Donald Nicholson-Smith. Oxford: Basil Blackwell.

Lemke, Jay. 1993. "Education, Cyberspace, and Change." *Electronic Journal of Virtual Culture* 1, no. 1.

Levidow, Les, and Kevin Robins. 1989. "Towards a Military Information Society?" In *Cyborg Worlds: The Military Information Society,* ed. Les Levidow and Kevin Robins. London: Free Association Books.

———. "Soldier, Cyborg, Citizen." 1995. In *Resisting the Virtual Life: The Culture and Politics of Information,* ed. James Brook and Iain A. Boal. San Francisco: City Lights Press.

Lindberg, David C. 1976. *Theories of Vision from Al-Kindi to Kepler.* Chicago: University of Chicago Press.

Lohr, Steve. 1998. "It Takes a Child to Raze a Village." *New York Times,* 5 March, D1.

Lovekin, David. 1989. "Technique and the Commonplace of the Commonplace." In *Commonplaces,* ed. David Black et al. Lanham, Md.: University Press of America.

Lukes, Steven. 1972. *Emile Durkheim: His Life and Work.* New York: Harper and Row.

Lyon, David. 1994. *The Electronic Eye: The Rise of Surveillance Society.* Minneapolis: University of Minnesota Press.

Lyotard, Jean-François. 1984. *The Postmodern Condition.* Trans. Geoff Bennington and Brian Massumi. Minneapolis: University of Minnesota Press.

MacKenzie, Donald. 1996. *Knowing Machines: Essays on Technical Knowledge.* Cambridge, Mass.: MIT Press.

Manovich, Lev. 1992. "Virtual Cave Dwellers: Siggraph '92." *Afterimage,* October.

———. 1995a. "An Archeology of a Computer Screen." In *NewMediaTopia.* Moscow: Soros Center for Contemporary Art.

———. 1995b. "To Lie and to Act: Potemkin's Villages, Cinema, and Telepresence." In *Ars Electronica* 1995 catalog. Linz, Austria: Ars Electronica.

Markley, Robert. 1996a. "Boundaries: Mathematics, Alienation, and the Metaphysics of Cyberspace." In *Virtual Realities and Their Discontents,* ed. Robert Markley. Baltimore, Md.: Johns Hopkins University Press.

———. 1996b. "Introduction: History, Theory, and Virtual Reality." In *Virtual Realities and Their Discontents,* ed. Robert Markley. Baltimore, Md.: Johns Hopkins University Press.

Marris, Peter. 1987. *Community Planning and Conceptions of Change.* New York: Routledge.

Marvin, Carolyn. 1988. *When Old Technologies Were New.* New York: Oxford University Press.

Marx, Karl. 1976. *Capital: A Critique of Political Economy.* Trans. David Fernbach. Vol. 1. Harmondsworth: Penguin.

Marx, Leo. 1965. *The Machine in the Garden: Technology and the Pastoral Ideal in America.* New York: Oxford University Press.

Mattelart, Armand. 1994. *Mapping World Communications: War, Progress, Culture.* Trans. Susan Emanuel and James A. Cohen. Minneapolis: University of Minnesota Press.

Mazlish, Bruce. 1967. "The Fourth Discontinuity." *Technology and Culture* 8, no. 1.

McCaffery, Larry. 1991. "Introduction: The Desert of the Real." In *Storming the Reality Studio,* ed. Larry McCaffery. Durham, N.C.: Duke University Press.

McCauley, Michael E., and Thomas J. Sharkey. 1992. "Cybersickness: Perception of Self-Motion in Virtual Environments." *Presence* 1, no. 3.

McKenna, Terence. 1991. *The Archaic Revival.* New York: Harper Collins.

McLuhan, Marshall. 1964. *Understanding Media.* New York: McGraw Hill.

Mecklermedia. 1995. *Final Program and Show Directory, Spring VRWORLD '95 Conference and Exhibition.* San Jose, Calif., 22–25 May.

Menser, Michael, and Stanley Aronowitz. 1996. "On Cultural Studies, Science, and Technology." In *Technoscience and Cyberculture,* ed. Stanley Aronowitz et al. New York: Routledge.

Meyrowitz, Joshua. 1985. *No Sense of Place.* New York: Oxford University Press.

————. 1993. "Images of Media: Hidden Ferment—and Harmony—in the Field." *Journal of Communication* 43, no. 2.

Miles, Ian, and Kevin Robins. 1992. "Making Sense of Information." In *Understanding Information: Business, Technology, and Geography*, ed. Kevin Robins. London: Belhaven Press.

Minsky, Margaret. 1984. "Manipulating Simulated Objects with Real-World Gestures Using a Force and Position Sensitive Screen." In *SIGGRAPH '84 Conference Proceedings*. New York: ACM.

Minsky, Margaret, et al. 1990. "Feeling and Seeing: Issues in Force Display." *Computer Graphics* 24, no. 2 (March).

Mitchell, Don. 1995. "There's No Such Thing as Culture: Towards a Reconceptualization of the Idea of Culture in Geography." *Transactions of the Institute of British Geographers* 20, no. 1.

Mitchell, William J. 1996. *City of Bits: Space, Place, and the Infobahn.* Cambridge, Mass.: MIT Press.

Mitchell, W. J. Thomas. 1994. "Imperial Landscape." In *Landscape and Power,* ed. W. J. T. Mitchell. Chicago: University of Chicago Press.

Morse, Margaret. 1990. "An Ontology of Everyday Distraction." In *Logics of Television,* ed. Patricia Mellencamp. Bloomington: Indiana University Press.

————. 1994. "What Do Cyborgs Eat? Oral Logic in an Information Society." In *Culture on the Brink: Ideologies of Technology,* ed. Gretchen Bender and Timothy Druckrey. Dia Center for the Arts, no. 9. Seattle: Bay Press.

Moulthrop, Stuart. 1993. "Writing Cyberspace: Literacy in the Age of Simulacra." In *Virtual Reality: Applications and Explorations,* ed. Alan Wexelblat. Cambridge, Mass.: Academic Press Professional.

Mulvey, Laura. 1975. "Visual Pleasure and Narrative Cinema." *Screen* 16, no. 3.

Mumford, Lewis. 1934. *Technics and Civilization.* New York: Harcourt Brace.

Nancy, Jean-Luc. 1991. *The Inoperative Community.* Trans. Peter Conner and Lisa Garbus. Minneapolis: University of Minnesota Press.

Nelson, Theodor H. 1972. "As We Will Think." In *International Conference on Online Interactive Computing, ONLINE 72 Conference Proceedings.* Uxbridge, U.K.: Brunel University.

————. 1973. "A Conceptual Framework for Man-Machine Everything." In *AFIPS Conference Proceedings,* no. 42. Washington, D.C.: Thompson.

Newton, Isaac. 1946. *Mathematical Principles of Natural Philosophy.* Ed. and trans. Florian Cajori. Rev. ed. Berkeley: University of California Press.

Novak, Marcos. 1992. "Liquid Architectures in Cyberspace." In *Cyberspace, First Steps,* ed. Michael Benedikt. Cambridge, Mass.: MIT Press.

Noyes, John. 1992. *Colonial Space: Spatiality in the Discourse of German South West Africa, 1884–1915.* Chur, Switzerland: Harwood.

Nunes, Joan. 1991. "Geographic Space as a Set of Concrete Geographical Entities." In *Cognitive and Linguistic Aspects of Geographic Space,* ed. David Mark and Andrew Frank. Boston: Kluwer Academic Publishers.

Nye, David E. 1994. *American Technological Sublime.* Cambridge, Mass.: MIT Press.

Oettermann, Stephan. 1997. *The Panorama: History of a Mass Medium.* Trans. Deborah Lucas Schneider. New York: Zone.

Olalquiaga, Celeste. 1992. *Megalopolis: Contemporary Cultural Sensibilities.* Minneapolis: University of Minnesota Press.

Ong, Walter J. 1977. *Interfaces of the Word.* Ithaca, N.Y.: Cornell University Press.

———. 1991. "Print, Space, and Closure." In *Communication in History: Technology, Culture, Society,* ed. David Crowley and Paul Heyer. White Plains, N.Y.: Longman.

Pausch, Randy, Thomas Crea, and Matthew Conway. 1992. "A Literature Survey for Virtual Environments: Military Flight Simulator Visual Systems and Simulator Sickness." *Presence* 1, no. 3.

Pesce, Mark D. 1993. "Final Amputation: Pathogenic Ontology in Cyberspace." Paper presented at Cyberconf3, University of Texas–Austin, 14–15 May.

Pickles, John. 1995. "Representations in an Electronic Age." In *Ground Truth: The Social Implications of Geographic Information Systems,* ed. John Pickles. New York: Guilford.

Piercy, Marge. 1991. *He, She, and It.* New York: Alfred A. Knopf.

Pinker, Steven. 1994. *The Language Instinct: How the Mind Creates Language.* New York: HarperPerennial.

Plato. 1987. *The Republic.* Trans. Desmond Lee. New York: Penguin.

Pollack, David. 1988. "The Creation and Repression of Cybernetic Man." *Clio* 18, no. 1.

Pomorska, Krystyna. 1984. Foreword to *Rabelais and His World,* by Mikhail Bakhtin. Trans. Hélène Iswolsky. Bloomington: Indiana University Press.

Porta, Giovanni Battista della. 1658. *Natural Magick.* London: Thomas Young and Samuel Speed.

Porush, David. 1996. "Hacking the Brainstem: Postmodern Metaphysics and Stephenson's *Snow Crash.*" In *Virtual Realities and Their Discontents,* ed. Robert Markley. Baltimore, Md.: Johns Hopkins University Press.

Poster, Mark. 1990. *The Mode of Information.* Chicago: University of Chicago Press.

Proffitt, Dennis R., and Mary K. Kaiser. 1991. "Perceiving Environmental Properties from Motion Information: Minimal Conditions." In *Pictorial Communication in Virtual and Real Environments*, ed. Stephen Ellis. New York: Taylor and Francis.

Quittner, Joshua. 1995. "Automata Non Grata." *Wired*, April.

Raulet, Gerard. 1991. "The New Utopia: Communication Technologies." *Telos* 24, no. 1.

Reese, William L. 1980. *Dictionary of Philosophy and Religion*. Atlantic Highlands, N.J.: Humanities Press.

Relph, Edward C. 1976. *Place and Placelessness*. London: Pion Press.

Rheingold, Howard. 1991. *Virtual Reality*. New York: Simon and Schuster.

Rodaway, Paul. 1994. *Sensuous Geographies: Body, Sense, and Place*. London: Routledge.

Romanyshyn, Robert D. 1989. *Technology as Symptom and Dream*. New York: Routledge.

Ronell, Avital. 1994. "Video/Television/Rodney King: Twelve Steps beyond the Pleasure Principle." In *Culture on the Brink: Ideologies of Technology*, ed. Gretchen Bender and Timothy Druckrey. Dia Center for the Arts, no. 9. Seattle: Bay Press.

Ross, Andrew. 1991. *Strange Weather: Culture, Science, and Technology in the Age of Limits*. New York: Verso.

Rothenberg, David. 1993. *Hand's End: Technology and the Limits of Nature*. Berkeley: University of California Press.

Rushkoff, Douglas. 1994. *Cyberia: Life in the Trenches of Hyperspace*. San Francisco: HarperCollins.

Ruskin, John. 1848. *Modern Painters*. New York: Wiley and Putnam.

Sack, Robert. 1980. *Conceptions of Space in Social Thought*. Minneapolis: University of Minnesota Press.

———. 1986. *Human Territoriality: Its Theory and History*. Cambridge: Cambridge University Press.

———. 1988. "The Consumer's World: Place as Context." *Annals of the Association of American Geographers* 78, no. 4.

———. 1992. *Place, Modernity, and the Consumer's World*. Baltimore, Md.: Johns Hopkins University Press.

Sass, Louis A. 1994. "Civilized Madness: Schizophrenia, Self-Consciousness, and the Modern Mind." *History of the Human Sciences* 7, no. 2.

Schroeder, Ralph. 1996. *Possible Worlds: The Social Dynamic of Virtual Reality Technology*. Boulder, Colo.: Westview Press.

Schulz, Jeffrey. 1993. "Virtu-Real Space: Information Technologies and the Politics of Consciousness." *Leonardo* 26, no. 5.

Schwartz, Robert. 1994. *Vision: Variations on Some Berkeleian Themes*. London: Blackwell.

Searle, John. 1995. *The Construction of Social Reality*. New York: Free Press.

Seltzer, Mark. 1992. *Bodies and Machines*. New York: Routledge.

Sennett, Richard. 1991. *The Conscience of the Eye: The Design and Social Life of Cities*. New York: Alfred A. Knopf.

Shapin, Steven, and Simon Schaffer. 1985. *Leviathan and the Air-Pump: Hobbes, Boyle, and the Experimental Life*. Princeton, N.J.: Princeton University Press.

Sheppard, Eric. 1993. "GIS and Society: Ideal and Reality." Position paper for NCGIA conference on "Geographical Information Systems and Society," Friday Harbor, Wash., 11 November.

Sherman, Barrie, and Philip Judkins. 1993. *Glimpses of Heaven, Visions of Hell: Virtual Reality and Its Implications*. London: Hodder and Stoughton.

Shields, Rob. 1991. *Places on the Margin: Alternative Geographies of Modernity*. London: Routledge.

Shilling, Chris. 1993. *The Body and Social Theory*. London: Sage.

Simpson, Lorenzo C. 1995. *Technology Time and the Conversations of Modernity*. New York: Routledge.

Smith, Neil, and Cindi Katz. 1993. "Grounding Metaphor: Towards a Spatialized Politics." In *Place and the Politics of Identity*, ed. Michael Keith and Steve Pile. London: Routledge.

Smith, Paul. 1988. *Discerning the Subject*. Minneapolis: University of Minnesota Press.

Stallybrass, Peter, and Allon White. 1986. *The Politics and Poetics of Transgression*. Ithaca, N.Y.: Cornell University Press.

Starobinski, Jean. 1989. *The Living Eye*. Trans. Arthur Goldhammer. Cambridge, Mass.: Harvard University Press.

Stenger, Nicole. 1992. "Mind Is a Leaking Rainbow." In *Cyberspace, First Steps*, ed. Michael Benedikt. Cambridge, Mass.: MIT Press.

Stephenson, Neal. 1992. *Snow Crash*. New York: Bantam Books.

Sterling, Bruce. 1993. "War Is Virtual Hell." *Wired* 1, no. 1.

Steuer, Jonathan. 1992. "Defining Virtual Reality: Dimensions Determining Telepresence." *Journal of Communications* 42, no. 4.

Stone, Allucquere Rosanne. 1992a. "Virtual Systems." In *Incorporations*, ed. Jonathan Crary and Sanford Kwinter. New York: Zone.

———. 1992b. "Will the Real Body Please Stand Up? Boundary Stories about Virtual Cultures." In *Cyberspace, First Steps*, ed. Michael Benedikt. Cambridge, Mass.: MIT Press.

Sun Microsystems Computer Corporation. 1994. *Virtual Reality from Sun*, July.

Sutherland, Ivan. 1963. "Sketchpad: A Man-Machine Graphical Communication System." In *AFIPS Conference Proceedings*, vol. 28. Washington, D.C.: Thompson.

———. 1965. "The Ultimate Display." In *Proceedings of the International Federation of Information Processing Societies Congress*, vol. 2. Amsterdam: North-Holland.

———. 1968. "A Head-Mounted Three Dimensional Display." In *AFIPS Conference Proceedings* 33, no. 1. Washington, D.C.: Thompson.

Synthetic Pleasures. 1996. Motion picture. 83 minutes. Color. Directed by Iara Lee. Produced by George Gund. Caipirinha Productions, New York.

Taylor, Charles. 1989. *Sources of the Self*. Cambridge, Mass.: Harvard University Press.

———. 1994. *Multiculturalism*. Princeton, N.J.: Princeton University Press.

Thuan, Trihn Xuan. 1995. *The Secret Melody: And Man Created the Universe*. Trans. Storm Dunlop. New York: Oxford University Press.

Tuan, Yi-Fu. 1974. *Topophilia: A Study of Environmental Perception, Attitudes, and Values*. Englewood Cliffs, N.J.: Prentice Hall.

———. 1977. *Space and Place: The Perspective of Experience*. Minneapolis: University of Minnesota Press.

———. 1978. "Sign and Metaphor." *Annals of the Association of American Geographers* 68, no. 3.

———. 1980. "Rootedness versus Sense of Place." *Landscape* 24, no. 1.

———. 1982. *Segmented Worlds and Self*. Minneapolis: University of Minnesota Press.

———. 1990. "Realism and Fantasy in Art, History, and Geography." *Annals of the Association of American Geographers* 80, no. 3.

———. 1993. *Passing Strange and Wonderful: Aesthetics, Nature, and Culture*. Washington, D.C.: Island Press.

Turing, Alan. 1950. "Computing Machinery and Intelligence." *Mind* 59, no. 236.

Turkle, Sherry. 1995. *Identity in the Age of the Internet*. New York: Simon and Schuster.

Vidler, Anthony. 1992. *The Architectural Uncanny*. Cambridge, Mass.: MIT Press.

Virilio, Paul. 1989. *War and Cinema*. Trans. Patrick Camiller. London: Verso.

———. 1994. "Cyberwar, God, and Television." Interview by Louise Wilson. Trans. G. Illien. *CTHEORY* 21 (October).

von Gierke, Henning E., and Eugene Steinmetz. 1961. *Motion Devices*. National Academy of Sciences—National Research Council Publication 903. Washington, D.C.

Walter, Eugene Victor. 1988. *Placeways*. Chapel Hill: University of North Carolina Press.

Wann, John, and Simon Rushton. 1994. "The Illusion of Self-Motion in Virtual Reality Environments." *Behavioral and Brain Sciences* 17, no. 2.

Wark, McKenzie. 1993. "Lost in Space: Into the Digital Image Labyrinth." *Continuum* 7, no. 1.

Weiner, Norbert. 1948. *Cybernetics, or Control and Communication in the Animal and the Machine*. New York: John Wiley.

———. 1989. *The Human Use of Human Beings: Cybernetics and Society*. London: Free Association Books.

Whalen, Terence. 1992. "The Future of a Commodity." *Science Fiction Studies* 19, part 1 (March).

Williams, Raymond. 1960. *The Long Revolution*. London: Chatto and Windus.

———. 1983. *Keywords: A Vocabulary of Culture and Society*. New York: Oxford University Press.

Williamson, Judith. 1978. *Decoding Advertisements*. London: Marion Boyars.

Winner, Langdon. 1993. "Social Constructivism: Opening the Black Box and Finding It Empty." *Science as Culture* 3, part 3, no. 16.

Woolley, Benjamin. 1992. *Virtual Worlds: A Journey in Hype and Hyperreality*. Oxford: Blackwell.

Index

Ken Hillis is assistant professor of communication studies at the University of North Carolina at Chapel Hill. He has a Ph.D. in human geography from the University of Wisconsin at Madison and has published articles in such journals as *Urban History Review, Cartographica, Ecumene,* and *Progress in Human Geography.*